Civil Engineering: The Science of Structures

Civil Engineering: The Science of Structures

Editor: Drew Morgan

NYRESEARCH
P R E S S

New York

Published by NY Research Press
118-35 Queens Blvd., Suite 400,
Forest Hills, NY 11375, USA
www.nyresearchpress.com

Civil Engineering: The Science of Structures
Edited by Drew Morgan

International Standard Book Number: 978-1-63238-592-5 (Hardback)

Cataloging-in-Publication Data

Civil engineering : the science of structures / edited by Drew Morgan.
    p. cm.
Includes bibliographical references and index.
ISBN 978-1-63238-592-5
1. Civil engineering. 2. Structural engineering. I. Morgan, Drew.
TA145 .C58 2018
624--dc23

Contents

Preface

Every book is a source of knowledge and this one is no exception. The idea that led to the conceptualization of this book was the fact that the world is advancing rapidly; which makes it crucial to document the progress in every field. I am aware that a lot of data is already available, yet, there is a lot more to learn. Hence, I accepted the responsibility of editing this book and contributing my knowledge to the community.

Civil engineering is concerned with the planning, development and design of buildings and physical structures. It consists of urban design and landscape architecture. Some of the major constructions that fall under this domain are dams, canals, roads, bridges, etc. This book elucidates new techniques of civil engineering and their applications in a multidisciplinary approach. It strives to provide a fair idea about this discipline and to help develop a better understanding of the latest advances within this field. It is meant for students who are looking for an elaborate reference text on civil engineering and its related fields.

While editing this book, I had multiple visions for it. Then I finally narrowed down to make every chapter a sole standing text explaining a particular topic, so that they can be used independently. However, the umbrella subject sinews them into a common theme. This makes the book a unique platform of knowledge.

I would like to give the major credit of this book to the experts from every corner of the world, who took the time to share their expertise with us. Also, I owe the completion of this book to the never-ending support of my family, who supported me throughout the project.

Editor

FLEXIBLE HOUSING: THE ROLE OF SPATIAL ORGANIZATION IN ACHIEVING FUNCTIONAL EFFICIENCY

Seyed Reza Hosseini Raviz, Ali Nik Eteghad, Ezequiel Uson Guardiola, and Antonio Armesto Aira
Department of Architectural Projects (DPA), Polytechnic University of Catalonia (UPC), Barcelona, Spain
seyedi.reza.hosseini.raviz@estudiant.upc.edu

....

Abstract
According to inhabitants' dimensions and various aspects of human life, flexible spaces are used as a solution in social housing due to the lack of space that architects always confront. In fact, flexible housing responds to inhabitants' needs throughout time. In other words, it evolves from the change in residents' requirements and promises adaptability to their living conditions. This spatial adaptability replaces spatial hierarchy and enhances life quality. The main purpose of spatial organization is to plan an interior space in order to create functional efficiency in a dwelling layout. This study attempts to research less-focused concepts so as to establish guidelines for future flexible housing design. Encompassing two case studies regarding Dutch housing, this study aims to understand how interrelated space planning enhances spatial arrangements to achieve an efficient spatial configuration. Spatial organization is delved with in depth to understand how functional efficiency can be achieved in flexible housing. Finally, interrelated spatial organization is believed to contribute to placing spaces according to their function by creating a multilateral relationship that responds to the inhabitants' ever-changing needs.

Keywords: *flexible housing; functional efficiency; spatial organization; flexibility.*

INTRODUCTION

Residential housing was the first form of architecture that people built for themselves. The primary structural behaviour of the first communities was nomadic or semi-nomadic and their settlements were temporary and movable. Once the first cores of the cities arise, the construction of housing as buildings began; later on, nevertheless, their basic characteristics changed only in inessential features (Förster, 2006). In fact the dwelling as a settlement, throughout the history of humanity, has been the basis of indigenous societies and cultural development. Housing is always influenced by living habits, cultural boundaries and environmental conditions. A house is designed based on the combination of comfort and safety and also a sense of belonging to a dwelling space (Mohamad Mahd Shabani, Tahir, Shabankareh, Arjmandi & Mazaheri, 2011).

Nowadays, architects not only focus on designing the facade composition and the geometry of the floor plan; they also take into consideration subjects such as the organization within and around the dwelling, the private and public space, the importance of spatial function and domestic structure, the gendered character of interior and exterior spaces, the influence of consumption patterns regarding spaces and decoration, the ways in which space is organized and many other aspects of the occupants' experiences (Lane, 2007). The main concern with respect to the conditions of the industrial cities was the low standards endured by many urban residences. All of the members of a family would usually inhabit a single room and share unsuitable bathrooms and kitchens with many other people. Even in the early philanthropic houses, where their undertaking and standards were generally good, kitchens and bathrooms were commonly shared. The main purpose of dwelling renovation in the twentieth century is to move from this situation and reach good space standards in housing design (Towers, 2005).

LITERATURE REVIEW

"The prognosis for proponents of modernist design and process was to generate flexibility in the design process. Hill (2003: 32) suggests that flexibility has many meaning nuances, but that it originally refers to the accommodation of changing relationships between events, contexts, and the use of space. The most common meaning of it is flexibility by technical means, which, as Forty (2000) states, can be understood in relation to a couple of types. The first one is flexibility by movement or the reconfiguration of the dwelling's intricate elements. The second type of technical flexibility refers to the use of lightweight demountable fixtures and fittings, and movable floors, walls and ceiling panels, including open-plan design. Flexibility by means of moveable parts and/or open plan is, for Hill (2003), as much a description of use as it is of form, and it is characterized by a versatile combination between space and use. The flexibility of a space depends in part on the user, or, as Hill (2003:38) suggests, the change of use may well be less dependent on a physical transformation of the space than a change in the perception of the user" (Imrie, 2006).

The concept of flexibility in the context of architectural housing is introduced under two topics: "the evolving conditions of the vernacular" and the "external pressures that have prompted housing designers and providers to develop alternative design solutions, including flexible housing" (Schneider & Till, 2007). According to this, it can be claimed that flexibility in domestic architecture either evolves and improves from the experience of traditional tendencies in housing design or appears as a new design tendency which follows the outward forces of the twentieth century (Albostan, 2009). For example, the main idea of spatial configuration employed in traditional Malay housing is associated with social and cultural patterns, and religious values. The most prominent feature of these houses is their spatial flexibility. The open plan with minimum physical boundaries offers flexibility to the space (Abdul Rahim & Abu Hassan, 2012). Additionally, flexibility is a feature of traditional courtyard housing which is an epitome of introverted structures in Iran (Nosratpour, 2012). In fact, creating different types of space for different functions implies that these houses intend to fulfil family requirements according to their lifestyle throughout time (Arjmandi, Tahir, Che-Ani, Abdullah, & Usman, 2010).

In fact, flexibility as a solution, in today's social housing, is an issue that has been considered on different levels throughout time in eastern architecture. In Japanese traditional architecture, sliding doors have the finality of separating spaces as well as changing the dimensions of these in the house. As a result, multi-functional spaces are created by opening them up. (Shabani, Tahir, Arjmandi, Abdullah & Usman, 2010).

For instance, in the rural house designed by Kazuhiko and Kaoru Obayashi in Osaka bay, "The actual flexibility and adaptability of the dwelling is completely dependent on the active participation of its users (as well as a specific type of furniture): by pulling out futons from a storage cupboard, a room that is used as a dining or sitting room can be transformed into a bedroom; the minimal use of furniture and the relative lack of untidiness demands discipline in order to achieve flexibility, which may be beyond normal living patterns, but nonetheless the principle remains. Flexibility is also enabled by means of a modular approach to design. The size of the rooms is based on the standard dimension of tatami mats; the house has rooms made up of a set of these mats. The openness of the plan as well as the frame construction suggest that functional and social changes can be dealt with easily - both on a daily basis as well as on a periodic or even longer term. Connections between rooms can be undertaken by opening or closing sliding screens, which can change the size and function of a space in a matter of seconds: two individual rooms can be united by simply opening up two large screens so that a couple of small spaces become one large room that can be used either for a specific festivity or for a family gathering" (Figure 1) (Schneider & Till, 2007).

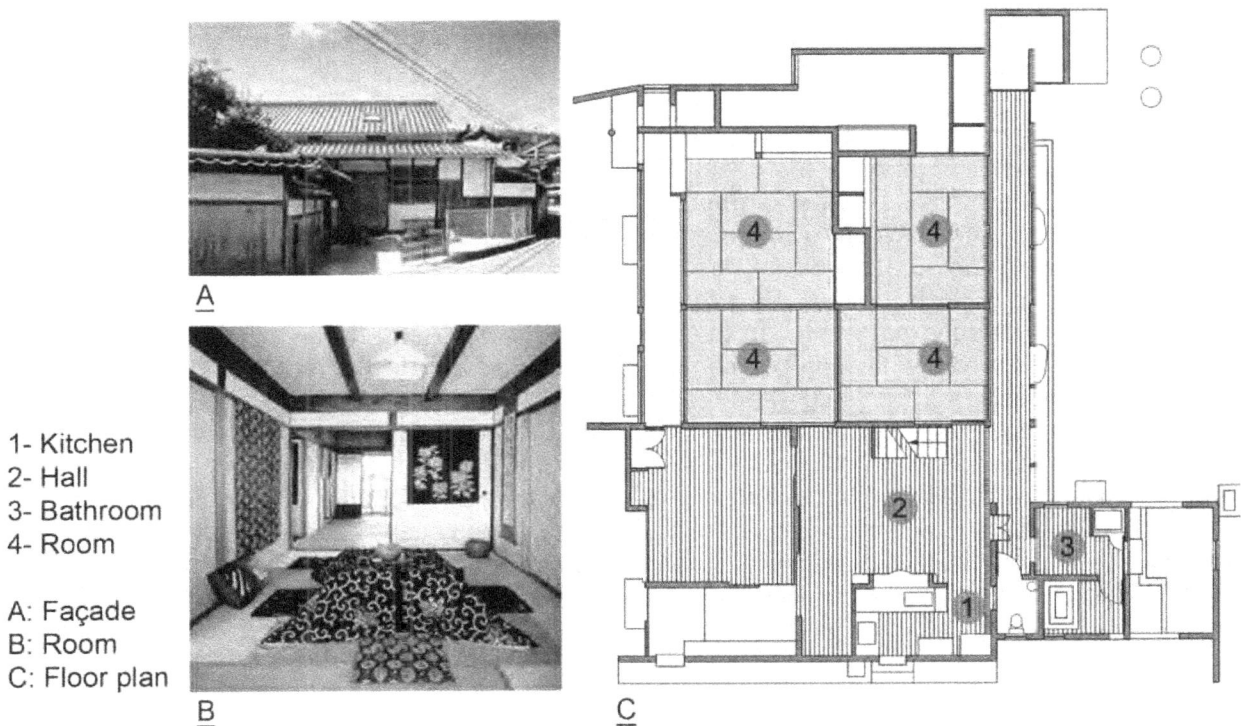

1- Kitchen
2- Hall
3- Bathroom
4- Room

A: Façade
B: Room
C: Floor plan

Figure 1. rural house designed by Kazuhiko and Kaoru Obayashi in Osaka bay
Source: Photos: Schneider, T., & Till, J. (2007). Flexible housing. Oxford: Architectural Press, Plan:
Authors

Designing Flexible floor plans has been experimented since 1920. Particularly the Netherlands has a long and ongoing tradition to which great architects such as Rietveld, Stam Van Doesburg, Van den Broek, Van Tijen, Habraken, Hertzberger and Van Eyck, and also the most recent generations, have made their innovative contributions. However, the spirit of flexibility appeared to flow throughout Europe in this century and the Netherlands was the country where this subject was systematically integrated in the construction development of social housing and where it has never completely disappeared from sight since then (Van Eldonk & Fassbinder, 1990).

RESEARCH METHODOLOGY
The manner in which interrelated space planning enhances spatial arrangements in order to achieve an efficient spatial configuration will be corroborated in this article. The main goal of this paper is to develop functional efficiency by means of improving spatial organization in flexible housing. This approach helps achieve adaptability between domestic space and life, and ultimately enhances the quality of life. The evaluation method is based on floor plan analysis encompassing two case studies of Dutch modern social housing; the Vroesenlaan apartment blocks designed by the Dutch architect J .H. van den Broek in the Blijdorp neighbourhood of Rotterdam (1934-35) and the Bergpolder apartment building designed by the Dutch architect Willem van Tijen in the north of Rotterdam (1933-34).

STATE OF THE MATTER
"At a basic level the case of flexible housing is a straightforward matter of common sense. Why, to put it simply, would one not design in terms of flexibility and adaptability? Housing is volatile, - subject to a whole range of cyclic, non-cyclic and tendency changes-, and if t is not able to respond to these changes it becomes at best unsatisfactory, at worst obsolescent. Yet, despite

the fact that a dwelling is inevitably dynamic, it is often framed intellectually and physically as something immobile. In fact, one of the problems of treating housing as if it were a static commodity with rigid design parameters is that it reaches a world of changing demographics. A mix of units that meet immediate demand might well be inappropriate in thirty, let alone one hundred, years' time. Thus, over the past twenty years there has been a decrease in the number of traditional family units, a higher proportion of elderly people, an increase in the number of single-person households, a rise in the demand for shared accommodation and a growing move towards home-working. Statistical data shows that these tendencies will probably continue into the next decades, though they will be overlapped by as yet unseen and uncertain demographic developments. Probably the only thing that one can state with any certainty is that the needs of housing at the end of the twenty-first century will be different to the requirements and desires today; the necessity for housing that can adapt to these changing demographics becomes compelling. Changing demographics require new architectural solutions that incorporate flexibility into emergent types of housing." Housing should also respond to the internal changes occurring during the lifetime of its inhabitants. These internal micro-changes come at a housing unit level. If it cannot adapt then the inhabitants will have to move home, which is both financially and socially disruptive. Housing has to be flexible enough to deal with two conditions. Firstly, the necessity to adapt to the changing needs of individuals as they grow old or become less physically capable. Secondly, a dwelling that can respond to the changing constitution of a family as it grows and then later on contracts (Schneider & Till, 2007).

Nowadays, as in the past, we build or change houses in order to respond to our requirements beyond it being a basic refuge. In turn, due to the fact that dwellings are a physical space, they impose limitations on their inhabitants and at the same time create opportunities for them (Ward, 1999). The evolution of the private habitat is an intrinsic feature of human beings. Housing is the means where inhabitants have the most environmental intervention power because it is the first factor of relationship with the environment. It is the place where inhabitants and their needs are encountered. Any alteration, change, creation or modification is an attempt to reach the sense of belonging to a place. A home is a place where family life occurs, as a series of changing or permanent conflicts. The heterogeneity of living life makes each house a customizable, unique and unrepeatable place (Valenzuela, 2004).

Architects now pay considerable attention to the users of residential units whose creative motivations have an influence on their homes. According to past experiences which have seldom been recognized, the inhabitants of a dwelling, including the owner or the family group who resides there, shape their homes. Sometimes they design and build them by themselves. At times they design them, and employ a carpenter, builder or architect. From time to time they influence the design by choosing out of a series of patterns. Nevertheless, inhabitants always modify and remake their home spaces by rebuilding, decorating, furnishing, remodelling, landscaping or simply by dwelling within the forms and spaces of domestic architecture. Therefore, the places where people inhabit are comprehended as a space in which to illustrate the ideas of family, individuality, lifestyle, privacy and socio-cultural patterns; to create, in other words, the general cultural patterns of an era (Lane, 2007).

Spatial organization influences the way in which family members find themselves inside a dwelling or within a building. The distribution and arrangement of space is a building aim and not just its physical objective. Put another way, buildings are not just objects, but transformations of space by means of objects. As a consequence, configuration is a fundamental relationship between form and space, which is appropriated in the processes, by which buildings are transformed from physical objects into social and cultural objects. In both senses, society acquires a definite and recognizable spatial order (Abdul Rahim & Abu Hassan, 2012)

"...Market research in the Netherlands has shown that people are more likely to stay in their homes if they can adapt them, and by a corollary high percentage want to move because they cannot adjust their dwellings to their needs" (Danko, 2013). Therefore, applying users' ideas in the planning and designing processes of the dwelling is a way to find out the relation between people's expected needs and their upcoming ones. This enhances the adaptability of the dwelling to the needs of its occupants and consequently their satisfaction. Providing adaptability and flexibility to dwelling spaces according to different lifestyles is a feature of the ideal home (Abbaszadeh, Kalani Moghadam & Saadatian, 2013)

Frank Bijdendijk, Director of the Amsterdam Housing Corporation, argues that without the love and pride of its inhabitants, a building is not assured a long life even when the flexibility requirement is met. To gain its users' love, a building need not necessarily meet standards which are popular among architects. However, it need not be in conflict with them either. "The meaning of what an architecture loved by its inhabitants is, is no doubt debatable, but the requirement needed unmistakably connects architecture to the quality of the everyday environment" (John Habraken, 2008).

Furthermore, "Flexible housing directly addresses issues of social and economic sustainability. Social aspects are not only sufficed by the user's involvement, but also by the capacity of flexible housing to accept demographic change and thus stabilise communities. The economic aspects are addressed by means of the long-term vision that flexible housing engenders through future-proofing and avoiding obsolescence. The beauty of flexible housing is that if one follows through its principles and combines them with a response to climate change, one almost inevitably reaches a sustainable solution that integrates the complete range of sustainable issues; however, the green rhetoric is a quiet one that eschews the superficial gestures of some types of sustainable architecture. Flexible housing potentially exceeds the accepted definition of sustainability — providing for the needs of the present without compromising the ability of future generations to meet their own needs — in as much as it is not about avoiding future compromise but encouraging the coming change" (Schneider & Till, 2007).

Flexible housing is aimed to conceive appropriate spaces for inhabitants with diverse lifestyles. In this manner, the ability to respond to users' demands, starting from the very beginning of the dwelling's occupation and lasting over time, can be considered as the main objective of flexibility in the context of domestic architecture. In other words, flexibility creates adaptable residential spaces according to the requirements of inhabitants with different lifestyles. Flexibility and adaptability, in this sense, are closely associated (Albostan, 2009). The primary intended scope of flexibility in domestic architecture is the creation of a building that can remain in use longer due to being able to satisfy current requirements rather than being used under an external drive. The advantages of flexibility include the capability to fulfil its objective in an ameliorated way by accommodating the occupant's intervention, accepting new technology and being more economically and ecologically viable. Accordingly, these buildings can respond to these alterations by adapting their use or function (Geehee Yu, 2011).

THE PURPOSE OF FLEXIBLE HOUSING

Flexible housing is identified as a planning choice in the design phase of domestic architecture; either both in terms of construction and social use, or designed for change over its lifetime. The degree of flexibility is highlighted in two ways. First of all the in-built opportunity for adaptability, defined as enabling different social uses, and second the opportunity for flexibility, defined as enabling different physical arrangements. The incorporation of flexibility into the design allows architects the illusion of designing their control over the building in the future, beyond the period of their actual responsibility for it (Schneider & Till, 2005).

In fact, flexible housing is a layout that can be adapted to inhabitants' requirements and will lead to carrying out their expectations and demands with their own collaboration. Inhabitants' collaboration in this process will finally result in increasing the general satisfaction of the users (Zandiyeh, Mehdi; Eghbali, Seyed rahman; Hesari, 2012). In general, a flexible house is a dwelling layout that can adapt to changing requirements and patterns, both social and technical issues. These changing demands may be technological (e.g. the updating of old services), practical (e.g. the onset of old age) or personal (e.g. an expanding family). The changing patterns might be demographic, environmental or economic ones. Hence, flexible housing undertakes all of the housing development process. In fact, flexibility in domestic architecture allows its' inhabitants to take part in the design process of the different possibilities of using their living space. Therefore, inhabitants have the opportunity to carry out adaptations to their home spaces (Schneider & Till, 2007).

Flexibility, as a helpful and effective method, has been utilized in different architecture spaces to reach functional efficiency. It has a comprehensive function in architecture that can be defined by open plans and sections, and by portable and changeable elements or furniture. Flexibility as an initial solution, in today's modern social housing, is a subject that has been employed in different levels throughout different time periods. The Dutch architect, Herman Hertzberger, said that when flexibility became the catchword, it was the panacea to cure all of the illnesses of architecture (M M Shabani et al., 2010).

CASE STUDY 1: VROESENLAAN HOUSING COMPLEX

The Vroesenlaan apartment blocks, designed by the Dutch architect J.H. van den Broek and built by the housing association De Eendracht in the Blijdorp neighbourhood of Rotterdam, have become a symbol of modern architecture (Komossa, 2005). J.H. van den Broek was one of the protagonists and forerunners of flexible design. He stated that by means of a more efficient spatial arrangement of the floor plan and by integrating folding beds and sliding walls, the housing dwelling typology could become smaller without losing comfort (Schneider & Till, 2007). In this housing complex, three characteristics of modern architecture have been incorporated: 1) An open block structure, in which one head end of the block is eliminated and the corners are opened up. 2) A modern spatial hierarchy that organizes the floor plan in order to achieve day and night use. 3) A facade design which was revolutionary for its time, with large glass areas and a concrete skeleton with yellow brick filling the open spaces. When designing these housing blocks, the architect replaced the closed urban block for three wings, arranged around a courtyard which is like an ornamental garden. The corners are open, omitting in this manner the corner flats with poor sun light. The U shape form created is open towards the direction of Vroesen Park, so that all of the dwellings have a view of Vroesen Park from the rear. The dwellings on the street side are 9.5 meters deep and 7.5 meters wide. All of the apartments have four rooms and the area of each dwelling is about 72 m² (Komossa, 2005). The floor plan was designed to respond to the family's changing needs. The small corridor, which is close to the entrance in the centre of the floor plan, has three doors. One gives access to the kitchen. The other two doors, which are next to each other and separated by a short stretch of wall between them, give access to a long space. The extended central space that functions as a living / dining and study room can be divided into two rooms of the same size by means of sliding doors (Figure 2). The room next to the kitchen is designated to be a dining and living room and the rest of the area is dedicated to a study room, although these functions may be changed throughout time. The study room can be converted into a bedroom by folding down beds that were incorporated into the design (Figure 3) (Schneider & Till, 2007).

1- Study room (flexible space)
2- Living/Dining room (flexible space)
3- Bedroom (flexible space)
4- Bedroom
5- Kitchen
6- Bathroom
7- Rest room
8- Hall
9- Balcony
10- Sliding doors

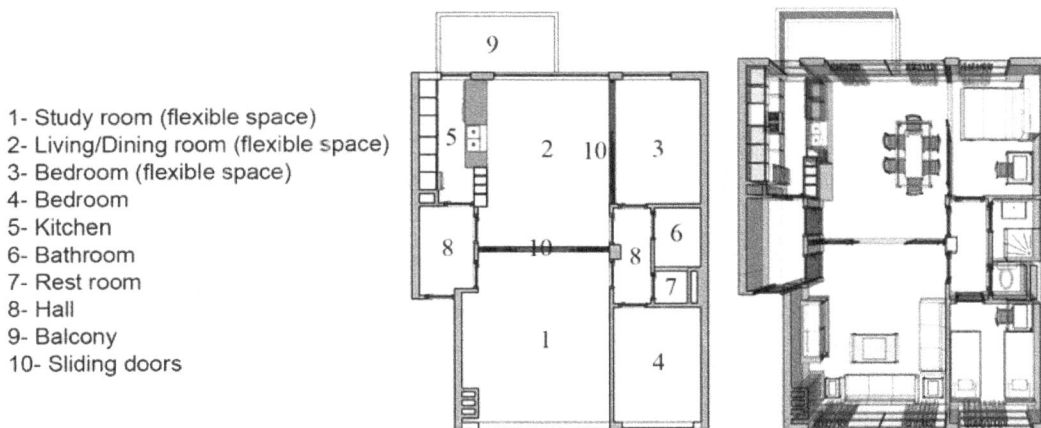

Figure 2. Plan view (Source: Author)

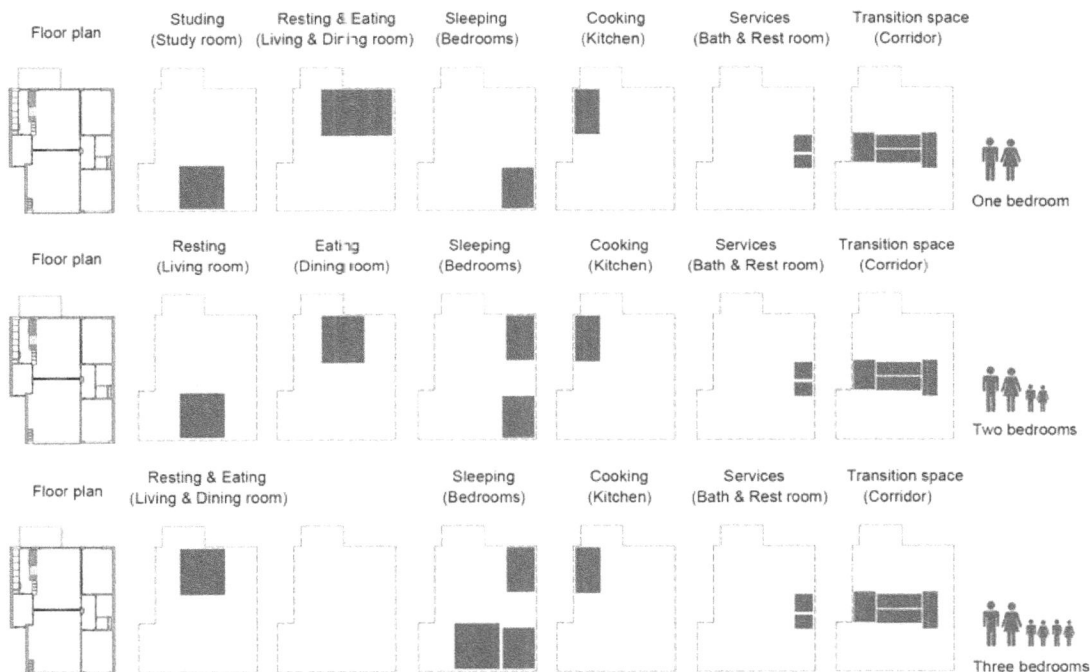

Figure 3. Vroesenlaan housing floor plan - Different spatial organization (Source: Authors)

On the other side of this extended space, there is a sequence of sliding doors that separate the living and dining room from a small room, one door, a short stretch of wall, another door (which mirrors the disposition on the opposite side of the room) and then a longer partition wall. Behind the two doors, there is another small corridor that has another four doors. One door opens to the room next to the dining and living room, the second one provides access to the bathroom, the third one leads onto the rest room, and the last one provides access to the only existing bedroom of the house (Schneider & Till, 2007).

Actually, the architect proposes day and night use, entirely in the analytic-functionalist tradition. Opening the sliding doors during the day and folding the parents' and older children's beds into built-in cabinets, providing an L shape living space with a sitting room and a playroom for the smaller children. The kitchen is designed to be efficient and minimal with a long Cabinet. It also has a door that opens onto the balcony, where the coal box is located (Komossa, 2005).

CASE STUDY 2: BERGPOLDER APARTMENT BUILDING

The Bergpolder apartment building was constructed between 1933 and 1934, it is considered a pioneer social housing block due to its height. The Bergpolder block is built in a working class area, in the north of Rotterdam (Martí Arís & Alegre, 1991).

The Dutch architect, Willem van Tijen, had a passion for designing empirical social housing. This building was one of the series of high-rise housing projects that Van Tijen designed in Rotterdam. The block includes 72 flats for small working-class households destined to couples or families with young children. It has nine floors with eight apartments per floor. They all follow the same type of floor plan design with an area of approximately 50 m². The spatial design of the Bergpolder flats is quite simple. Each floor is formed by four pairs of mirror-like flats which are accessed via a gallery (Barbieri, van Duin, de Jong, van Wesemael, & Wilms Floet, 2003).

The architect carried out a careful study in order to achieve efficient space in the flats. He wanted the flats to be practical and to create a roomy space sensation in them in spite of their small size. He also adopted the principles of a day and night floor plan. The flats dimensions are 8.2m deep and 6.2m wide, and have a clear height of 2.55m (Figure 4). Its spatial organization has been designed as follows: acceding from the gallery, one enters a corridor in which on one side is the kitchen and on the other side is the rest room as well as the bathroom. The corridor connects the entrance to the living room, which has a store cupboard on one side and on the other side a door that opens onto the children's bedroom (Figure 5). The master bedroom and the living room adjoin the balcony. They are kept separate from each other by a glass sliding door. During the day, the beds swing up into the wall and the sliding doors are opened (Barbieri et al., 2003).

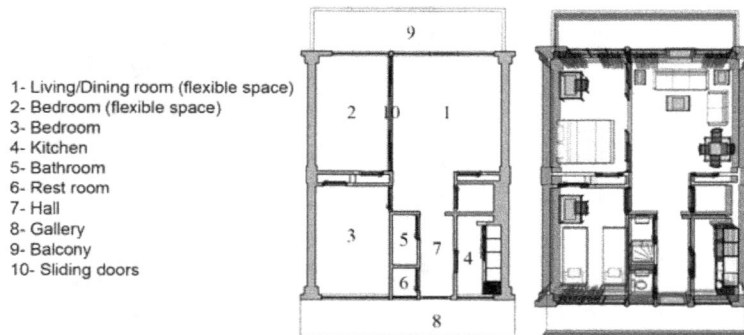

1- Living/Dining room (flexible space)
2- Bedroom (flexible space)
3- Bedroom
4- Kitchen
5- Bathroom
6- Rest room
7- Hall
8- Gallery
9- Balcony
10- Sliding doors

Figure 4. Plan view (Source: Authors)

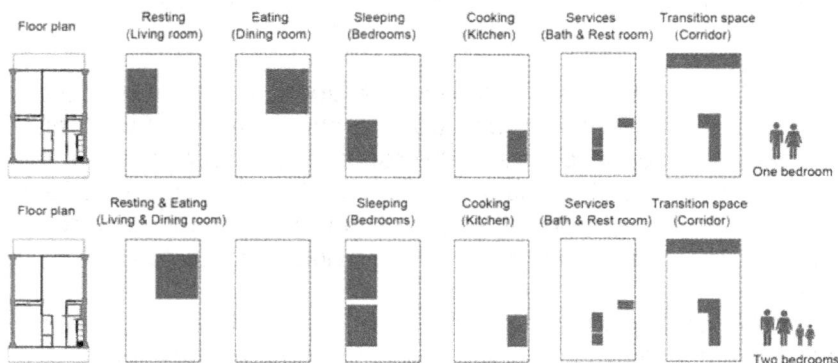

Figure 5. Bergpolder apartment floor plan - Different spatial organization (Source: Authors)

Glass sliding doors and folding beds create the possibility to use the flats in two different ways. During the daytime, the beds are folded up and the master bedroom joins the living room space; and the glass sliding doors are closed at night time ("Architecture in Rotterdam" n.d.). In fact, this separation via sliding doors allows improving the function and size of the living area throughout the daytime. Furthermore, the generated space, with a long dimension which is parallel to the windows of the balcony, creates a pleasant living room during the daytime (Gringberg & Bakema, 1977).

DISCUSSION

In order to develop the idea of "responding to changing needs" in housing, Van den Broek collaborated with H. Leppla, who undertook the research concerning apartments' requirements. Leppla carried out detailed studies of the processes of night-time and daytime uses, according to various family members' lifestyles. He related studies of day and night zones to the studies of the life cycles of the different household members and to their requirements and changing customs. In this work, he distinguished the different life phases that could perfectly occur in the course of a family's life cycle (baby, child, son, daughter, husband and wife). A dwelling, he argued, had to be able to meet all the functional needs of these individual users (van Eldonk & Fassbinder, 1990).

Following the consideration that the organization of architectural space is principal, the apartments' fix spaces and elements as well as its flexible ones have been arranged (Figure 6). The dimension and spatial distribution of both apartments can be easily changed according to the occupants' requirements during the day and night-time (Figure 7). In fact, not only is there a spatial independency but there is also a multilateral cooperation between architectural spaces and the liberty of choosing among the different options provided to the occupants throughout the day and night-time. As a result, the inhabitants can make multi-functional spaces, with minor changes, in both apartments. In the Vroesenlaan housing complex, the restroom and bathroom are integrated as a compact part of the layout while allowing considerable freedom to the remaining part of it. The restroom unit does not have any connection whatsoever to the kitchen and the open kitchen is connected to the dining room and a flexible bedroom. Together they can create an enormous space in the day time. Therefore, the flexible bedroom and the living room are integrated with each other so as to make a suitable space for the day time.

Fixed space	Changeable space	Perspective	Fixed space	Changeable space	Perspective

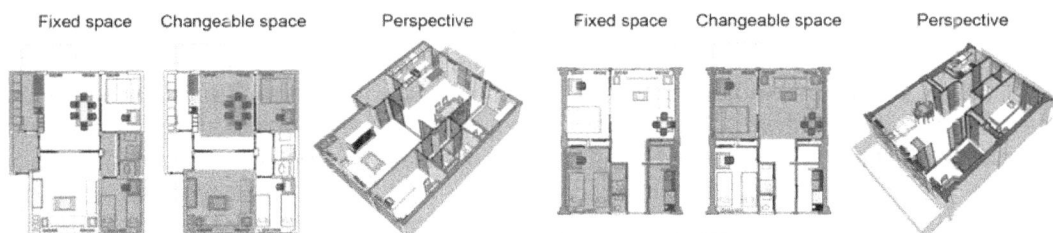

Figure 6. Fixed and changing spaces. Left: Vroesenlaan apartment. Right: Bergpolder apartment (Source: Authors)

In both apartments, there is at least one bedroom which is defined as a multifunction space. These bedrooms are not accessible directly from the hallway and have ultimately given the inhabitants the opportunity to be able to extend their space for daily activities. The kitchen is one of the most important spaces in housing design. It is not only for cooking and serving foods but it can also be used as a dining or living room or a place where the family can get together and do their diverse domestic activities. The activities that take place in these spaces have a close connection. As an outcome, their spatial layout must be in an adjacent connection, which consequently leads to reforming spatial hierarchy. In both projects, the architects have considered the kitchen and bathroom units as static spaces. In other words, they are placed on the floor plan as fixed-feature spaces. The living/dining room and one bedroom work as multifunction spaces. Space has been divided and organized based on spatial hierarchy by making appropriate connections according to its inhabitants' needs, activities and lifestyles. In fact, organizing the interrelated flexible spaces has enabled its inhabitants' freedom of choice. Home spaces have the capability to be arranged and configured according to inhabitants' lifestyles and changing requirements. Put another way, each family member has the autonomy to engage various activities in complete freedom. In both apartments, the organization of interrelated flexible spaces gives inhabitants the liberty to change their living spaces and helps achieve functional efficiency in housing.

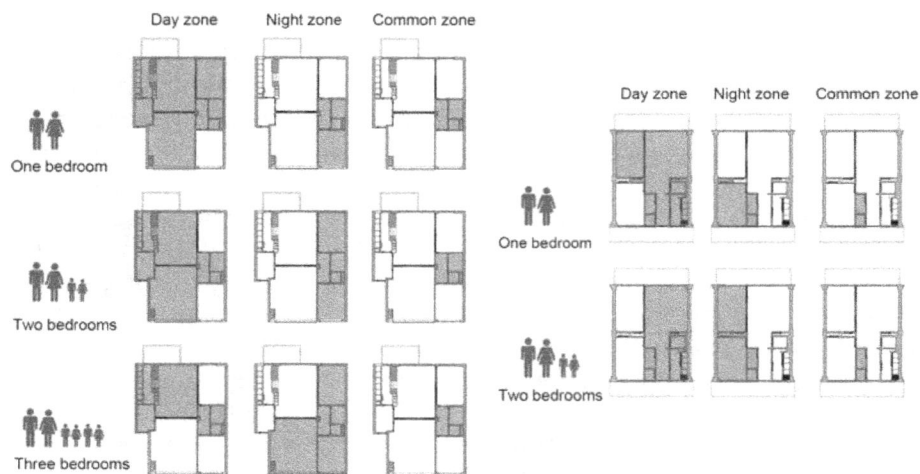

Figure 7. Day and night time spaces according to different spatial organization. Left:Vroesenlaan apartment. Right: Bergpolder apartment (Source: Authors)

CONCLUSION

Flexible housing presents an opportunity for inhabitants to participate in the design of their dwelling and to arrange their living spaces according to their lifestyles and needs by creating new and temporary spaces during the day and night time. Based on both presented case studies, architectural spaces may be subject to change in order to meet inhabitants' requirements. This entails the autonomy of incorporating various activities when necessary and enhancing the variability and versatility of the connections between adjacent spaces without any geometrical change in the form of the architectural spaces. In both of the apartments' layouts, family members have been offered the opportunity to redesign and rearrange their household in accordance to their changing lifestyles. The dimensions of the rooms can vary when conforming new spaces and fulfilling new requirements without extra expenditure. If the number of house members increases, the dwelling's layout will have the possibility to change to a new layout. In fact, the flexibility of space in housing seeks to create diversity, dynamism and adaptability which are decisive in order to satisfy inhabitants' variable needs. Spatial organization determines the boundaries of the functions of space in housing. Hence, it's main task being that of arranging spaces according to its residents' requirements. The spatial organization of the floor plans of both of the cases studied is based on the interaction between different parts of the dwelling, which are configured as far as their function is concerned by situating spaces in close connection. Sequential and interrelated spaces create new spaces with various types of spatial relationships by employing sliding doors, walls and flexible elements. These new spaces meet the needs of the dwelling's residents, which arise from the activities of the family members during day/night-time. This spatial organization creates interaction between its domestic spaces so as to achieve functional efficiency. In an overall view, flexible housing gives its inhabitants the opportunity to get involved in the design process of creating a suitable spatial environment throughout their lifetime. This will notably present them with a sense of belonging to their living place by fulfilling their expectations as well as by adapting it to their different demands instead of taking an architecturally-predetermined approach. Flexible housing places its emphasis on the fact that advanced architecture not only is put forth through form, but is accomplished in practice by discreetly incorporating functionality and usability into the dwelling's layout with the aim to accommodate the diverse needs of its inhabitants over a long period of time.

REFERENCES

Abbaszadeh, S., Kalani Moghadam M., & Saadatian, O. (2013). ANALYZING A PROPER FLEXIBLE AND ADAPTABLE PATTERN FOR PROMOTING THE HOUSING QUALITY IN IRAN. *Design + Built*, 6.

Abdul Rahim, A., & Abu Hassan. F. (2012). Study on Space Configuration and Its Effect on Privacy Provision in Traditional Malay and Iranian Courtyard House. *International Proceedings of Economics Development & Research* (Vol. 42, pp. 115–119).

Albostan, D. (2009). *FLEXIBILITY IN MULTI-RESIDENTIAL HOUSING PROJECTS: THREE INNOVATIVE CASES FROM TURKEY*. The Graduate School of Natural and Applied Sciences of the Middle East Technical University.

Architecture in Rotterdam. (n.d.). Acquired on December 20, 2013, from http://www.architectuurinrotterdam.nl/building.php?buildingid=248&lang=en&PHPSESSID=228f4fae1 3d7250ef550110c6208b879

Arjmandi, H., Tahir, M. ., Che-Ani, A ., Abdullah, N. A. ., & Usman, I. M. . (2010). Application of Transparency to Increase Day-Lighting Level of Interior Spaces of Dwellings in Tehran - A Lesson from the Past. *Selected topics in power systems and remote sensing*, 297–307.

Barbieri, U., van Duin, L., de Jong, J., van Wesemael, P., & Wilms Floet, W. (2003). *A Hundred years of Dutch architecture :1901-2000 : trends, highlights* (p. 370). Amsterdam: Sun.

Danko, M. (2013). *Designing Affordable Housing for Adaptability: Principles, Practices, & Application*. Claremont Colleges.

Van Eldonk, J., & Fassbinder, H. (1990). *Flexible fixation :the paradox of Dutch housing architecture = De paradox van de Nederlandse woningbouw* (p. 83). Assen: Van Gorcum.

Förster, W. (2006). *Housing in the 20th and 21st centuries =Wohnen im 20. und 21. Jahrhundert* (p. 175). München etc.: Prestel.

Geehee Yu, J. (2011). *FLEXIBLE URBAN HOUSING FOR CHANGING LIFESTYLES*. University of Maryland.

Gringberg, D. I., & Bakema, J. B. (1977). *Housing in the Netherlands :1900-1940* (p. 144). Delft: Delft University Press.

Imrie, R. (2006). *Accessible housing : quality, disability and design*. London [etc.] : Routledge.

John Habraken, N. (2008). Design for flexibility. *BUILDING RESEARCH & INFORMATION*, *36*(3), 290–296.

Komossa, S. (2005). *Atlas of the Dutch urban block* (p. 283). Bussum: Thoth.

Lane, B. M. (2007). *Housing and dwelling :perspectives on modern domestic architecture* (p. 467). Abingdon: Routledge.

Martí Arís, C., & Alegre, L. (1991). *Las Formas de la residencia en la ciudad moderna : vivienda y ciudad en la Europa de entreguerras* (p. 206). Barcelona: Servicio de Publicaciones de la UPC.

Nosratpour, D. (2012). Evaluation of Traditional Iranian Houses and Match it with Modern Housing. *Journal of Basic and Applied Scientific Research*, 2(3), 2204–2213.

Schneider, T., & Till, J. (2005). Flexible housing: opportunities and limits. *arq: Architectural Research Quarterly*, 9(2), 157–166.

Schneider, T., & Till, J. (2007). *Flexible housing* (p. 237). Oxford: Architectural Press.

Shabani, M. M., Tahir, M. M., Arjmandi, H., Abdullah, N. A. G., & Usman, I. M. S. (2010). Achieving Privacy in the Iranian Contemporary Compact Apartment Through Flexible Design. *Selected topics in power systems and remote sensing*, 285–296.

Shabani, M. M., Tahir, M. M., Shabankareh, H., Arjmandi, H., & Mazaheri, F. (2011). RELATION OF CULTURAL AND SOCIAL ATTRIBUTES IN DWELLING, RESPONDING TO PRIVACY IN IRANIAN TRADITIONAL HOUSE. *Journal of Social Sciences and Humanities*, 6(2), 273–287.

Towers, G. (2005). *An Introduction to urban housing design :at home in the city* (p. 316). Oxford etc.: Architectural Press.

Valenzuela, C. (2004). Plantas transformables La vivienda colectiva como objeto de intervención. *ARQ Ensayos y documentos Essays and documents (Santiago)*, n58, 74–77.

Ward, P. (1999). *A History of domestic space :privacy and the Canadian home*. Vancouver; Toronto: UBC Press.

Zandiyeh, Mehdi; Eghbali, Seyed rahman; Hesari, pedram. (2012). The Approaches towards Designing Flexible Housing. *Naghsh jahan*, 95–106.

AUTHORS

Seyed Reza Hosseini Raviz
PhD Student
Department of Architectural Projects (DPA)
Polytechnic University of Catalonia (UPC)
seyedi.reza.hosseini.raviz@estudiant.upc.edu

Ali Nik Eteghad
PhD Student
Department of Architectural Projects (DPA)
Polytechnic University of Catalonia (UPC)
ali.nik.eteghad@estudiant.upc.edu

Ezequiel Uson Guardiola
Professor
Department of Architectural Projects (DPA)
Polytechnic University of Catalonia (UPC)
ezequiel.uson@upc.edu

Antonio Armesto Aira
Professor
Department of Architectural Projects (DPA)
Polytechnic University of Catalonia (UPC)
antonio.armesto@upc.edu

2

ALLURE OF THE "CRYSTAL: MYTHS AND METAPHORS IN ARCHITECTURAL MORPHOGENESIS

Emine Özen Eyüce
Bahcesehir University, İstanbul

ozen.eyuce@arc.bahcesehir.edu.tr

Abstract

'Form' has always been one of the most important issues in architectural design. In the process of form-giving to the end-product, architects make use of different sources from typologies to intuitions or metaphorical ones. When the generic ideas of the prominent examples in architectural history have been traced, it can clearly be stated that one of the most effective metaphors used in architecture is the 'Crystal.' Appearing at the intersections between nature and human history and having a long history going back to myths, the 'Crystal' has been used extensively in architecture both as reflecting the meaning originating from its mythical background, and also, as a metaphor representing the perfection in nature. This article will try to trace the use of 'crystal' metaphor in history and analyzing the two examples, namely, the Royal Ontario Museum: 'Crystal' in Toronto (2007) by Daniel Libeskind and Musée des Confluences: 'Crystal Cloud of Culture' in Lyon (2014) by Coop Himmelb(l)au, will try to evaluate the change in the use of crystal metaphor in contemporary architectural morphogenesis.

Keywords: *Crystal; Myth; Metaphor; Nature; Morphogenesis*

INTRODUCTION

Crystal is one of the most sophisticated inorganic structures of nature which is repeatedly used in the past and still continues to be a significant source in the architectural form-giving process. An overview of the history reveals that crystal and crystalline formations have been used both literally and metaphorically in all arts and architecture from Antiquity up until today. The meanings attributed to crystals and the way they were incorporated into design process have varied in different periods of architectural history. They have come forward either as a reflection of a myth or as a symbolic metaphorical relationship representing transformations in social life and technology taking its cues from 'nature'. This timeless allure of the 'crystal' has been accepted as coming from its symbolism. It is the symbol of 'perfection', 'purity' and 'clarity' in the form derived from inorganic nature; the symbol of transformation from 'life to death', and also, the symbol of the ' organic' or 'living' with its potential to generate new forms of life. In contemporary architectural practice, crystal metaphor still continues to be an important symbol of natural processes in the morphogenesis of architectural end product. In other words, nature', once a source of inspiration for imitation by analogy, has become a source with its inherent principles discovered by scientific developments for the genesis of architectural form through computer based parametric design generations.

The Genealogy of Crystal Metaphor

Peter Behrens (1868-1940) explained the meaning of crystal as:"the symbolism of the crystal relies on a metaphorical relationship between transformations which take place at the micro- and macro- cosmic levels; for example, just as mere carbon under intense conditions assumes a particular crystal structure and becomes the prized diamond, so the power of art may transform everyday life into a resplendent life filled with meaning" (Anderson, 2002, p. 50). In this respect, he used crystal symbol for the opening ceremony of Exhibition for Darmstadt Artists' Colony in

1901. Although it was not an architectural example, the crystal symbol was used in the meaning of "metamorphosis of everyday life into a heightened artistic experience" (Bletter, 1981, p. 31). The most celebrated architectural example of 'crystal' in history is Crystal Palace (1851) which is accepted as "...a revolution in architecture from which a new style will date" (Giedion, 1967, p. 251), and in Lothar Bucher's words "all materiality blend into atmosphere" as cited in Giedion's famous book Space, Time and Architecture (Giedion, 1967, p. 255). Although named as crystal, the idea in the generation of from in Crystal Palace was not derived from the crystal form and idiosyncrasies of crystalline formations; instead, the technological advancements of the age used in garden structures of the period - iron skeleton and glass cladding were more effective in design. Paxton used the features of glass as 'transparency', 'lightness' and 'the ease of rationality in production' in his design. Crystal Palace was a real technological breakthrough in linear tectonics and the rationalized production techniques of the period, but, Ruskin criticized Paxton's design as "the New Crystal Palace as the poetical public insists upon calling it, though it is neither a palace nor of crystal...." (Ruskin, 2009, p. 96). In fact, these qualities, 'transparency and lightness' which were the prime motivation behind the use of glass in Crystal Palace were also the reason behind many glasshouses built at the end of the Seventeenth Century without any reference to mythical or metaphorical background. Glass became a favourite building material with its clarity and quantity of light in the age of enlightenment and rationalism instead of 'the aura of mysticism obtained by stained glass' in Gothic interiors (Hisham, 2006, p. 8). In the Nineteenth Century, glass developed to be an indispensible complementary building material in iron skeleton systems in glasshouses due to the developments in the glass industry such as Jardin de Plantes (1833), Munich Glass Palace (1834), Palm House in Kew Garden (1844).

The second recurrence of the crystal and crystalline forms as a metaphor appeared at the beginning of the Twentieth Century among German architects who sought to create an architecture free from traditional norms and constraints of the Nineteenth Century Schinkel tradition within utopian expectations in search for a new society against the political turmoil and social upheaval after the First World War. The idea of new architecture had expanded among architects belonging to a circle named as Crystal Chain (Die Gläserne Kette), generally referred as 'Utopian Correspondence' initiated by Bruno Taut, who signed his letters to friends with the pseudonym Glas. Taut designed Glass Pavilion for Werkbund Exhibition in Cologne just before the outbreak of First World War. Affected by his ideas about glass architecture, Taut dedicated this building to his friend, poet Paul Scheerbart, who is known for his book Glasarchitektur (1914) written with the inspiration under the effect of the glasshouse of the Dahlem Botanical Gardens in Berlin. The Glass Pavilion reflecting the ideals of the new architecture became the symbol of German Expressionism after the War. Taut, had linked the domed shape of his glass Pavilion with crystalline forms in his explanatory report as: "The large dome that resembles rhombohedron of crystal in its form, is composed of glass planes between ferroconcrete ribs and rests on iron-reinforced support, which comes out from a concrete scale" (Yamini-Hamedani, 2009, p.101). Rhombohedron crystal form was also used by Violet-le-Duc, in one of his drawings in which he juxtapose rhombohedrons of granite crystals with hexagonal crystals of volcanic basalt (Donahue, 1995, s. 50).

Taut's statement 'the Gothic Cathedral is the prelude to glass architecture' and one of the Scheerbart's couplets written on the plinth as "Light permeated the Universe/ It comes to life in crystal" (Scheerbart, 1972, p. 14) reflected Taut's interest both in mythical and spiritual meaning embodied in 'glass' architecture and also in debates of the Nineteenth-Century about the nature as a source of inspiration in architecture. What Scheerbart and Taut hoped for from 'Crystal Culture' was a new morality for the society (Pehnt, 1973, p. 74). Adolf Behne, who had hailed 'the alluring beauty of the ideal' - 'the ideal of glass architecture,' saw a delightful fragment of his ideal in the pavilion built for the glass industry (Pehnt, 1973, p. 75). He wrote in a 1915 review of Bruno Taut's architectural projects: "The longing for purity and clarity, for glowing lightness, crystalline exactness, for immaterial lightness, and infinite liveliness found in glass a means of its fulfilment—in this most bodiless, most elementary, most flexible, material" (Bletter, 1981, p. 34). Crystal, representing the ultimate and the supreme, became the symbol for this new architecture, named as Expressionist Architecture by Behne in 1915. This architecture, contrary to the Paxton's Crystal Place,"...has no other purpose than to be beautiful" according to the pamphlet prepared by Taut for the visitors of the Glasshaus (Ersoy, 2007, p. 240).

In search of the background of German Expressionist Architecture appearing at the beginning of the Twentieth Century, Rosemary Bletter traces the mythical origin of glass back to King Solomon referring to Old Testament, Biblical descriptions, and Koran. King Solomon is said to have built a palace of glass (with glass floors). When Queen of Sheba, not knowing with the illusory effects of glass architecture, upon entering Solomon's palace: (Bletter, 1981, p. 23) "....when she saw it, she thought it was a body of water and uncovered her shins [to wade through]. He said, "Indeed, it is a palace [whose floor is] made smooth with glass." She said, "My Lord, indeed I have wronged myself, and I submit with Solomon to Allah, Lord of the worlds" (Koran 27:44).

According to Bletter, the meaning of glass architecture and its suggestion of shimmering water is quite direct and literal in Solomonic Legend (Bletter, 1981, p. 25). On the contrary, both glass and crystal, which has been used interchangeably, represented a spiritual meaning and superiority as a result of their symbolic metaphorical consideration in later examples.

The Crystal metaphor reappeared in German Expressionist architecture, coincided with a paradigm shift 'from the history-based approaches of the Nineteenth Century into the Twentieth Century visions of abstract' (Inceköse, 2006, p. 10). Especially, following the publishing of Wilhelm Worringer's book 'Abstraction and Empathy' (1907) and later 'Problems Of Form in Gothic Art' in 1911, architectural discourse centered around the idea of abstract in art and architecture and the 'crystal' form was accepted as true reflection of the natural order as Gothic architecture did. It is stated in his book Words and Buildings by Adrian Forty that "for most of the last five hundred years 'nature' has been the main, if not the principle category for organizing thought about what architecture is or might be" (Forty, 2012, p. 220).

Although addressed from different perspectives, or in some periods, the superiority of human being over nature has been accepted as a result of worldview, nature has been a creative repertoire in the formation of built environment from Antiquity up today. In the Eighteenth Century, architects interested in the processes and the rules of the nature for imitating as the origins of built form, like Leon Battista Alberti, in search for the organizing rules in the harmony of parts, as expressed in his book De Re Aedificatoria (mid 15th C), Vitrivius's myth of first building in De architectura, or Abbe Laugier's Primitive Hut (1753).

Nature was one of the prime concerns in art and philosophy also in the Nineteenth Century. Together with an increasing interest in Goethe's natural history studies, Goethe's interests in both crystal formation and plant 'morphology' influenced not only the German architectural theory but also philosophy. In his book, The World As Will And Idea (1851), Schopenhauer, referring many times to Goethe, asserted the role of crystal in the formation of life and also its unity: ".......the crystal has only one manifestation of life, namely its formation, which afterwards has its fully adequate and exhaustive expression in the coagulated form, in the corpse of that momentary life" (Schopenhauer, 1969, p. 155) and also: "in the inorganic kingdom of nature all individuality completely disappears. Only the crystal can still to some extent be regarded as individual; it is a unity of the tendency in definite directions, arrested by coagulation, which makes the trace of this tendency permanent. At the same time, it is an aggregate from its central form, bound into unity by an Idea" (Schopenhauer, 1969, p. 155).

In the Nineteenth Century, as a result of scientific studies of mineralogy and history of the natural world's own formation- Geohistory, an accumulation of a new knowledge on both forms of the earth's surface and also of the underlying unities of the diverse forms of nature like crystals and plants affected all arts and also the architectural discourse. The architectural discourse's increasing interest in natural sciences with the tendency toward implanting theories, concepts and methods derived from natural sciences in architectural form-giving process reflected itself in the unconventional arguments in seminal works of three figures' at the beginning of Nineteenth Century: John Ruskin's (1819-1900) The Seven Lamps of Beauty (1949) and Stones of Venice (1851), Eugenie Viollet-le-Duc's (1814-1879) Dictionnaire Raissoné de l'architecture francaise (1856), and Gotfried Semper's (1803-1879) Der Stil (1860). Although they considered nature from different points of view, all three architects referred to crystal and crystalline forms. Semper, referring to geological formations using crystal metaphor for clarity and homogeneity stated that:"...just as the splendid marble that gives shape to the coasts and cliffs of Greece - notwithstanding its homogeneous formation- betrays its sedimentary origin through veins, scattered fossils, and other signs embedded in it, Hellenic art cannot deny its secondary origin. It too reveals to the observer all the deposits that form its material base but that, in a great metamorphosis of a whole people, rushed together from their sedimentary conditions into a crystal-clear homogeneity" (Bergdoll, 2007). In a similar way, fascinated with the Alps, both Ruskin and Viollet-le-Duc undertook extensive studies of the mineralogy and geology of the Alpine earth formations. Ruskin's exquisite watercolours of 'Fragment of the Alps' and by Viollet-le-Duc's studies of the glaciers of the High Alps outside Lausanne (Bergdoll, 2007) influenced art and architecture extensively that can be traced in Caspar David Friedrich's Sea of Ice (or Arctic Shipwreck) (1823-4) and in Walter Gropius's Memorial to March victims in Weimar (1921). (Figure1)

Figure 1.(Left) C.D.Friedrich's Sea of Ice, (1823-24) (http://www.wga.hu/art/f/friedric/3/309fried.jpg); (Right) Walter Gropius; Memorial to March Victims in Weimar (1921). (https://www.pinterest.com/pin/452048881324350949/)

Both Viollet-le-Duc and Ruskin sought analogies between natural formations and architecture especially with Gothic. Ruskin objecting the straight line's being at odds with nature, he discusses:"to find right lines in nature at all we may be compelled to do violence to her finished work, break through the sculptured and colored surfaces of crags, and examine the processes of their crystallization" (Donahue, 1995, s. 50).
In Stones of Venice, Ruskin stated that "(but) against crystalline form, which is the completely systematized natural structure of the earth... The four-sided pyramid, perhaps the most frequent of all natural crystals, is called in architecture a dogtooth; its use is quite limitless, and always beautiful ... and all mouldings of the middle Gothic are little more than representations of the canaliculated crystals of the beryl, and such other minerals" (Ruskin, 2009, p. 226).
Bruno Taut's famous book, named as Alpine Architecture, illustrating his ideals for a utopian future also explains the continuity of thoughts on nature behind German Expressionist Architecture. Although their utopian designs for glass buildings in 1920's couldn't find the

opportunity to be realized, the crystal has become the symbol of Bauhaus, most significant architectural movement in history, influenced by the ideas of Taut and Glass Chain. In the program pamphlet for opening speech in 1919, Walter Gropius explained the aim of the regeneration of German visual culture through the synthesis of arts and crafts as: "Let us create a new guild for craftsmen, without the class distinctions which raise an arrogant barrier between craftsman and artist. Together let us conceive and create the new building of the future, which will embrace architecture and sculpture and painting in one unity and which will rise one day toward heaven from the hands of a million workers like the crystal symbol of the new faith" (Curtis, 1996, p. 184). The cover of the program illustrated by Lyonel Feininger included a woodcut expressing a crystal cathedral, an allegory of the total work of art which represented the three arts of painting, sculpture and architecture as a symbol of social unity (Droste, 1993, s. 19). To summarize, a direct and literal use of glass in Solomonic myths has been transformed into a crystal metaphor for a new social order and salvation for architects. However, Bauhaus drew away from the complex and subtle ideas of Expressionist architecture fulfilling the demands of industrial productions and the functionalist approaches in design within the socio-political context of the age

The Crystalline forms faded from view during the political shifts of the early 1920's. Siegfried Gideon evaluated German Expressionism as: "The Expressionist influence could not perform any service for architecture" in his famous book Space, Time and Architecture (Giedion, 1967, p.485). On the contrary, parallel to the form priority approaches of 80's, as a reflection of the search for perfect form and geometry, crystal metaphor reappeared in the form-giving process. The Crystal Cathedral (1980) which has been designed as a religious monument with an appearance of a transparent four-pointed crystal by Philip Johnson and John Burgee, explained as hinted by the Bauhaus Manifesto illustrated by Feininger's woodcut: "the crystal symbol of a new faith."

When the latest examples of the architectural practice are analyzed, it is clearly seen that the Twenty-first Century has inherited the growing fascination for the crystal metaphor both in naming and in the visual appearance of the end-products. UFA Cinema Center in Dresden (1998) by Coop Himmelb(l)au; Denver Art Museum (2006), massive crystalline addition to existing Royal Ontario Museum (Michael Lee-Chin Crystal) in Toronto (2007) and Crystals at City Center, in Las Vegas (2009) by Daniel Libeskind; The Basque Health Department Headquarters by Coll-Barreu Arquitects in Bilbao (2008), and Musée des Confluences in Lyon by Coop Himmelblau (2014) ; and Soyak Crystal Tower in İstanbul are only some of the prominent examples that make use of the 'crystal metaphor'. All these examples bring to mind the following question: are the expressionist design approaches accepted as 'crystal utopias' at the beginning of the Twentieth Century coming to real with the help of technological advancements as in farsighted words of Otto Kohtz as early as 1909: "It s highly possible that later generations will achieve such mastery of materials and technique that they will construct a building or a landscape for no other purpose than that of contemplation, simply out of a desire to create in a particular mood, rather in the way that many pieces of music are written today" (Pehnt, 1973, p. 9).

Contemporary Use of Crystal Metaphor in Architectural Morphogenesis
Following the exhibition of Expressionist Utopias: Paradise, Metropolis, Architectural Fantasy in Los Angeles in 1993, the Crystal metaphor has emerged back from the memory, and the stunning design of the exhibition by Wolf D. Prix provoked architects' attention into Expressionist ideals of the 1920's.

In 1998, Coop Himmelb(l)au architectural office of Wolf D. Prix designed The UFA Cinema Center in Dresden. The UFA has displayed similarities with the visionary design ideas of the German Expressionism. The building, characterized by two intricately interconnected units: The Cinema Block– with eight cinemas and the Crystal- a glass shell which serves simultaneously as the foyer and public square, has been expected to be "a crystalline lamp displaying a series of complex and fragmented images to the city" (Heathcote, 2001). The contrast of lightness and sparkling brilliance of huge walls of glass and the heavy concrete structures containing the auditoria which anchor the building to the earth reinforce this notion of the revival of Expressionist imagery, of the building as a crystal cathedral rising mystically from the solid rock of the earth (Heathcote, 2001, p. 91) (Figure 2).

Figure 2. (Left) The UFA Cinema Center, Dresden The Crystal from outside; (Right) The Crystal from inside (Devrim Işıkkaya, 2012).

Coinciding with the Deconstructivism in architectural discourse, the attractive and unusual geometries proliferated at the beginning of the Twentieth Century. Libeskind designed a massive expansion to the Royal Ontario Museum: ROM, in Toronto. Named as 'The Crystal', the museum is simply an assemblage of five giant cubes in between two historic stone buildings. Thinking about the museums' role in rebuilding and revitalizing cities, Libeskind designed ROM with an unusual geometry to create public attraction and activity point. The 3-Dimensional intersection of predefined regular forms creates an enclosure for 'a multilevel space created at the intersection between the crystals' (Stanwick, 2007) in which the reclaimed natural light through the refracted spears piercing the cubes create a spacious effect. In ROM, The Crystal, on the contrary to the precedents of crystal metaphor, don't carry the characteristics of illuminating its environment with its gleaming crystal's "purity and clarity" and "infinite liveliness found in glass" as Behne stated in defining German Expressionism (Figure 3).

Figure 3. The Royal Ontario Museum:ROM, in Toronto (2007) by Daniel Libeskind

On the other hand, both The UFA and The Crystal- ROM are the prominent examples of a changing form paradigm that started to challenge the everlasting use of Cartesian space understanding in the form-giving process.

Today, a new paradigm of form as well as the idea of space developed in architecture as a result of the advances in digital technologies and modelling techniques. Paralleled with the changing environmentalist attitudes to the relations between human beings and nature, stimulated the alternative approaches, 'employing techniques and processes outside the mainstream of industrial production' (Forty, 2012, p. 238), 'nature' has become a new source of architectural quality. As a result of incorporating the volumetric matrices, growth, adaptation, pattern derivations as in nature, a shift in form paradigm transformed the architectural design from a form-giving act into a designing of a form-finding process. The new architecture, freed from Cartesian space understanding with the help of digital design processes, therefore, has also changed the idea of form. The current technological advances allow architects to use computers and digital tools for generating forms, in other words, morphogenesis. The concept of morphogenesis, originating from biology in the Nineteenth Century and transferred to the Twentieth Century by geology, is understood as a group of methods that employ digital media not as representational tools for visualization but as generative tools for the derivation of form and its transformation. Digital morphogenesis in architecture bears a largely analogous or metaphoric relationship to the processes of morphogenesis in nature, sharing with it the reliance on gradual development but not necessarily adopting or referring to the actual mechanisms of growth or adaptation (Roudavski, 2009, p. 348). In morphogenetic processes of design, instead of fitting into a predetermined form as in form-priority design paradigm, designers develop a generative model which proposes many alternatives for the designer to chose. Therefore, design process becomes a form-finding process among the numberless variability: "The plan no longer 'generates' the design; sections attain a purely analytical role. Grids, repetitions, and symmetries lose their past raison d'etre, as infinite variability becomes as feasible as modularity, and as mass-customization presents alternatives to mass-production" (Kolarevic, 2005, p. 13). The computational derivations out of nature have made possible even large scale complex structural productions like the Bird's Nest (2008), Water Cube (2007), and CCTV Headquarters Building (2008) in Beijing. All these projects taking inspiration from nature have created a new understanding of form. They also became the examples of mass-customized production operating through the logic of optimization with the use of digital morphogenesis.

Digital morphogenesis, or in other words, the new form paradigm as form-finding has also been applied to the designs with Crystal metaphor. In this new approach, crystalline aesthetic has developed to be meaning fractures, reflections, imperfections as in natural processes instead of accepting the beauty of a predefined perfect crystal form, parallel to the change in the evaluation of nature as a process. Nature and natural processes and crystals are still the main sources of inspiration in design. Deleuze and Guattari affected the architectural discourse as: "We took as our point of departure cases of this kind on the geological stratum, the crystalline stratum, and physicochemical strata, wherever the molar can be said to express microscopic molecular interactions ("the crystal is the macroscopic expression of a microscopic structure"; the "crystalline form expresses certain atomic or molecular characteristics of the constituent chemical categories") (Deleuze, 1987, p. 57).

Always in search for dynamic space, fluid context, vibrant representation of life, and vitality, Austrian Firm Coophimmelb(l)au designed Musée des Confluences (2014) in Lyon. The Museum is one of the contemporary examples of the Crystal metaphor in architecture that also makes use of the morphogenetic process in design. The Museum, taking its name from its site known as the "Pointe du Confluent," has been built on an old industrial area at the intersection of the Rhone and Saone rivers with the purpose of revitalizing the devastated industrial area. Taking its cue from its special site, the Confluence, Museum becomes a point for encounters between natural and man-made nature, science and art, education and recreation to fulfill its founding 'mission of increasing and disseminating knowledge among mankind.' (Couturier, 2014, p. 25) The building embodies three heterogeneous components as the Crystal, the Cloud, and the Plinth, bringing the earth and sky together within "progressive differentiation" in Manuel DeLanda's terms without homogenizing the parts (DeLanda, 2002, p. 14). Displaying the organic unity in the form of assemblages, Museum is expected to be a "Crystal Cloud of Culture" for the City of Lyon. (Figure 4)

Figure 4.(left) "The Crystal Cloud of Culture": Lyon; (Right) Entrance Hall, (© Duccio Malagamba, 2014) Musée des Confluences (2014) by Coop Himmelb(l)au.

The Crystal, housing the entrance hall, brings together the museum and the city and its immediate environs as an urban forum and connect entrance hall to exhibition spaces through the vertical circulation space. Entrance hall's crystalline form is completely structured by glass and steel. Wolf D. Prix explains the essence of the design as the 'concept of fluidity' that is derived from turbulence created by the flow of the two rivers confluence at the tip of the land. The flow starts from the river, continues into the entrance hall hosting the multi-faceted crystal structure that is constructed out of rectangular steel tubular frames carrying the transparent glass panels. Crystal's multifaceted planar surfaces supported by steel and glass structure transforms at one point into a curvilinear vortex touching the ground, as a supporting element called 'gravity well' (Figure 5).

By virtue of digital computational techniques, transformation of the 2-dimensional planar surface's rational grid structure into a 3-dimensional grid to create a support displays the morphogenetic character of the design process. The Cloud, hovering above the ground contains the exhibition spaces as Black Boxes. A grid system also continues within the Cloud creating multifaceted geometries, covered with metal sheets in contrast to the transparency of the Crystal's glass-panelled surfaces. As a result, the emergence of a form generating multiple reflections and visions both from the inside and outside, day and night, adds to the vitality of the city, similar to the Expressionist ideals. In fact, Wolf D. Prix has criticized the Expressionist architecture as "representing an insufficient level of formal achievement", although he has accepted influence on their design approach (Benson, Dimendberg, Frisby, Heller, & Kæs, 2001, p. 180). Departing from the regular forms with the help of digital technologies creating asymmetric, dynamic, multifaceted three-dimensional forms, crystal metaphor in contemporary architecture has become the symbol of three-dimensional dynamization of space. Adding 'movement' with the inclusion of human being into space, the crystal has been transformed from inorganic to organic living organism (Figure 6).

Figure 5. The Gravity Well: Entrance Hall, Musée des Confluences (2014) by Coophimmelb(l)au, Lyon (© Duccio Malagamba, 2014).

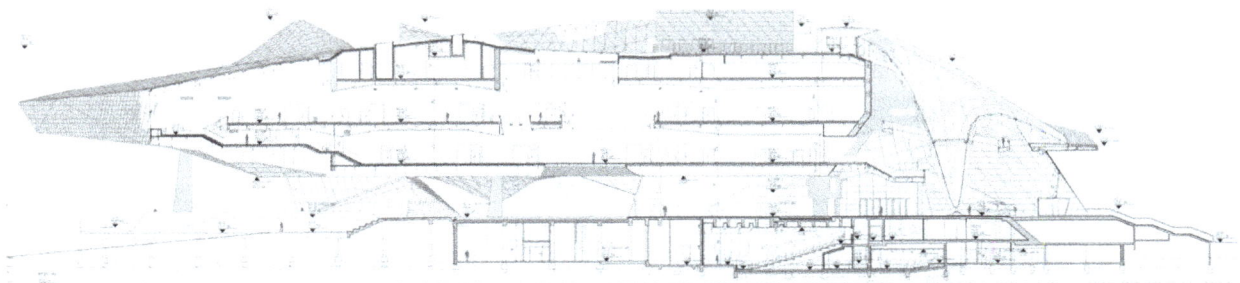

Figure 6. Musée des Confluences, Section (2014), (© Coop Himmelb(l)au).

Conclusion

This paper aimed at evaluating the transformation of the crystal metaphor from glass- a transparent and lightweight building material used as a cladding, to an architectural form representing the esoteric roots and the idea of perfectness, and finally the state-of-the-art approaches as crystalline formations in form-finding processes. Crystal metaphor has always been an important generic idea in design, either as a representation of a myth or as a metaphor reflecting the idea of perfectness, purity, and transformation derived from the 'nature' that is a source of inspiration either for imitation or driving the laws of order. The use of crystal metaphor traces a discontinuous emergence at the intersections between humanity- natural history and architectural thought, where the 'nature' has been accepted to develop human experience in creating arts. Since "mere imitation of natural forms and objects would achieve..... inferior and derivative beauty" as Ruskin stated, rejecting imitation advocated the "acceptance of sources of delight from nature" in the power of mental expression in architecture.

The confluence of humanity, nature and architecture also coincide with the intersections of art and natural sciences in history like geohistory, mineralogy where the ordering laws of nature discovered. In the contemporary world, harnessing the digital technologies in the generative process, a whole new world of formally and spatially different, fluid, dynamic digital designs, other than Cartesian space, has developed.(Figure 7)

Unfortunately, many architects incorporated natural forms from biology or zoology just for the sake of visual similarities to nature. On the other hand, the experimentation with natural forms and processes are the topic of many articles in recent years. Roudavski gives a detailed information on the studies related to natural morphogenesis in architecture. He discusses about the difference of the biomimetics and the bioinspiration which emphasizes "indirect and multiplicious characteristic of knowledge transfer between biology to architecture" (Roudavski, 2009, pp. 365-368). Most of the digital morphogenesis studies have concentrated on material performance over appearance, and on processes over representation (Leach, 2009, p. 34).

The reuse of patterns and tessellations inspired from the patterns of nature applied to many architectural designs like pinwheel aperiodic tiling of Federation Square (2002), Serpentine Pavilion by Toyo Ito or Voronoi tessellations are not only mere skins but also spatial designs.

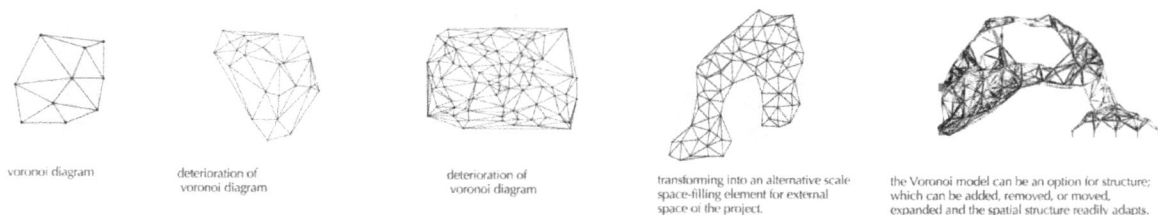

voronoi diagram   deterioration of   deterioration of   transforming into an alternative scale   the Voronoi model can be an option for structure;
                         voronoi diagram   voronoi diagram   space-filling element for external   which can be added, removed, or moved,
                                                                              space of the project.   expanded and the spatial structure readily adapts.

Figure 7. Voronoi Tessellations and alternatives for masses for an Aviation Museum Project by Öykü Arda, Graduation Project at Bahcesehir University (Source: Author).

Furthermore, modelling three dimensional forms using polygonal meshes, usually made up of triangles and quadrilaterals or deformed meshes with straight lines has ended in multifaceted continuous surfaces and spatial continuities as well. The new modeling technique making use of curves –NURBS, instead of straight lines together with the developments in evolutionary biology used in digital design and fabrication paved the way for more natural, adaptive and transformative architectural morphogenesis.

The Crystal's structural system in Musée des Confluences shows how the rectangular mesh transforms into a curvilinear surface creating a flow of outer space into interior creating a structural element. Therefore, differentiating from other architectural examples still accepting the formal analogy of crystal, Musée des Confluences shows the development one step further approaching to natural morphogenesis in the 'becoming' process.

REFERENCES

Anderson, S. (2002). *Peter Behrens and A new Architecture for Twentieth Century.* Cambridge, Massachusetts: The MIT Press.

Bärnreuther, A., Kahlfeldt, P., Kleihues, J. P., & Scheer, T. (2000). *City of Architecture, Architecture of the City: Berlin 1900- 2000.* Berlin: Nicolai Verlag.

Bergdoll, B. (2007). Of Crystals, Cells, and Strata: Natural History and Debates on the Form of a New Architecture in the Nineteenth Century. *50*, 1-29. Retrieved from http://www.jstor.org/stable/40333846.

Bletter, R. H. (1981). The Interpretation of the Glass Dream-Expressionist Architecture and the History of the Crystal Metaphor. *The Journal of the Society of Architectural Historians, 40* (1), pp. 20-43. http://doi.org/10.2307/989612.

Cheetham, M. A. (2010, June). The Crystal Interface in Contemporary Art:Metaphors of the Organic and Inorganic. *Leonardo, 43* (3), pp. 250-256.Retrieved from http://www.jstor.org/stable/40661659.

Couturier, H. L. (2014). *A great Story of Mankind: Musée des Confluences.* (D. v.-J. Pignard, Ed.) Lyon, France: Musée des Confluences.

Curtis, W. J. (1996). *Modern Architecture since 1900.* London: Phaidon .

DeLanda, M. (2002). Intensive Science and Virtual Philosophy. London: Continuum.

Deleuze, G. (1987). *Thousand Plateaus: Capitalism and Schizophrenia.* Minneapolis, MN, USA: University of Minnesota Press.

Donahue, N. H. (1995). *Invisible Cathedrals:The Expressionist Art History of Wilhelm Worringer.* (ed., Dü.) Pennsylvania: University of Pensylvania.

Droste, M. (1993). *Bauhaus 1919-1933.* Berlin: Taschen.

Ersoy, U. (2007, December). The Fictive Quality of Glass. *Architecture Research Quarterly , 11* (3-4), pp. 237 - 243.DOI: http://dx.doi.org/10.1017/S1359135500000737 .

Forty, A. (2012). Words and Buildings: A Vocabulary of Modern Architecture. Thames and Hudson.

Giedion, S. (1967). *Space Time and Architecture* (5th Edition ed.). Cambridge, Massachusetts: Harward University Press.

Hisham, E. (2006). *Cultures of Glass Architecture.* Burlington: Ashgate.

İnceköse, Ü. (2006). Instrumentalisation of Natural Sciences for The Reconstruction of Architectural Knowledge: Lissitzky, Doesburg, Meyer, Teige. *Unpublished Ph.D. Thesis Institute of Technology.* İzmir, Turkey.

Kolarevic, B. (2005). Digital Morphogenesis, In Architecture In The Digital Age: Design And Manufacturing. In B. Kolarevic (Ed.). Abingdon, Oxon: Taylor and Francis.

Kolarevic, B., & Klinger, K. (2008). Manufacturing/Material/Effect. In B. Kolarevic, & K. Klinger (Eds.), *Manufacturing Material Effects: Rethinking Design and Making in Architecture* (pp. 5-24). New York: Routledge.

Leach, N. (2009, January/February). Digital Morphogenesis. *Architectural Review, 79* (1), pp. 32-37.DOI: 10.1002/ad.806.

Pehnt, W. (1973). *Expressionist Architecture.* New York: Praeger Publishers.

Roudavski, S. (2009). Toward a Morphogenesis in Architecture. *International Journal of Architectural Computing , 03* (07), pp. 345-374. DOI: http://dx.doi.org/10.1260/147807709789621266.

Ruskin, J. (2009). The Stones of Venice, Volume II (of 3), December 31, [eBook #30755],p.96).

Scheerbart, P. (1972). *Glass Architecture.* (D. Sharp, Ed., & J. Palmes, Trans.) New York, USA: Praeger Publishers.

Schopenhauer, A. (1969). *The World as Will and Representation.* (E.F.J. Payne, Trans.) Dover Publ.

Shaw, D. B. (2008). *Technoculture.* New York: Berg Publishers.

Stanwick, S. (2007). The Crystal at the Royal Ontario Museum. *Architectural Design, 77,* pp. 126-127.

Taut, B. (1972). *Alpine Architecture.* (D. Sharp, Ed., & S. Palmer, Trans.) New York, USA: Praeger Publishers.

Timothy O. Benson, E. D. (2001). *Expressionist Utopias: Paradise, Metropolis, Architectural Fantasy.* Los Angeles, California: Los Angeles County Museum of Art.

Yamini-Hamedani, A. (2009). Nietzche's Zarathustra: Between Persia and Greece. In L. M. Roberts (Ed.), *Germany and Imagined East* (pp. 113-124). Newcastle: Cambridge Scholars Publishing.

EMINE ÖZEN EYÜCE

Emine Özen Eyüce
Instructor, Assoc.Prof. Dr.
Bahcesehir University,Faculty of Architecture
ozen.eyuce@arc.bahcesehir.edu.tr

3

ICT-ENABLED BOTTOM-UP ARCHITECTURAL DESIGN

Burak Pak
University of Leuven, Faculty of Architecture
Brussels, Belgium

*burak.pak@kuleuven.be

Abstract

This paper aims at discussing the potentials of bottom-up design practices in relation to the latest developments in Information and Communication Technologies (ICT) by making an in-depth review of inaugural cases. The first part of the study involves a literature study and the elaboration of basic strategies from the case study. The second part reframes the existing ICT tools and strategies and elaborates on their potentials to support the modes of participation performed in these cases. As a result, by distilling the created knowledge, the study reveals the potentials of novel modes of ICT-enabled design participation which exploit a set of collective action tools to support sustainable ways of self-organization and bottom-up design. The final part explains the relevance of these with solid examples and presents a hypothetical case for future implementation. The paper concludes with a brief reflection on the implications of the findings for the future of architectural design education.

Keywords: *Bottom-up; Participation; Architectural Design; Crowdsourcing; Crowdfunding; Self-organization.*

INTRODUCTION

In his seminal book "The City Shaped", Kostof (1991, p.43) identifies two different kinds of cities in history: planned and unplanned. The first one, *ville créée* refers to an urban pattern designed by an authority in a top-down manner. A clear example from the middle ages is the *nieuwestad* (new city) Naarden in the Netherlands with a grid plan dating from 1350.

The second kind is a city which grows dominantly from the bottom-up in a spontaneous manner: *ville spontanée*. It emerges as a result of unplanned evolution without a master plan, as a result of centuries of daily struggles and bottom-up spatial interventions of the citizens. Numerous cities fit in this description such as Cappadocia in the Göreme Valley, Turkey, dating from 1800 B.C. and Thera, Santorini in Greece, 3000 B.C (Rudofsky, 1964, p.58). In this context, bottom-up participatory architectural practices can hardly be considered as novel.

For thousands of years, architecture was an evolving and emergent communal work, in other words, a spontaneous and continuous activity of people with a common heritage, *acting under a community of experience* (Belluschi, 2012). Vernacular architecture was produced through the collective work of ordinary inhabitants through the use of local materials. Their simplicity and harmony with the environment made them sustainable.

Indeed, some examples with antecedents dating from 6000 B.C. are still occupied and functional today (Figure 1). Instead of trying to challenge the nature, these practices welcomed and made use of the climate. However, the most powerful affordance of the vernacular architecture was the *direct and unself-conscious translation into the physical form of a culture, its needs and values, as well as the desires, dreams, and passions of a people* (Rapoport, 1969, p.2).

Figure 1. Architecture without architects: Self-built vernacular Houses in Harran, Sanliurfa, Turkey (Source: Author).

Overall it is clear that vernacular architecture examples around the world have close associations with durability, versatility, flexibility, adaptability accompanied with a strong sense of community, identity, and place (Rudofsky, 1964, p.13).

However, as a result of the professionalization of architectural design beginning with the Renaissance as well as the industrialization in the following centuries, these essential priorities started to be replaced by cost-cutting mass-production approaches. Besides the loss of essential characteristics of spaces, *modernity* caused a significant shift in the understanding of the relationship between time and space. This transformation was in the form of the rejection of contingency *through the assumption of a state of perfection to be reached tomorrow* (Bauman, 2000, p.25, p.29). It was this moment when space and time became separated in the minds of the producers. The communal building aspects of architecture and collective endowment were deteriorated in favor of massive individualization and privatization of nearly all aspects of life. The architecture serving to its users' needs and values was no more.

As a reaction to the developments above, participatory bottom-up architectural practices were brought onto the agenda of architectural design prominently after the Second World War and gained traction with the movements of 1968 (Jencks, 2011). The following decades witnessed the emergence of alternative approaches to architectural design. A bottom-up pluralist strand became popular in Western Europe, the Americas and many other regions around the world as an alternative to state-centered and top-down models (Bowns and da Silva, 2011).

In this paper, I will make a brief review of these approaches with examples from specific cases which can be claimed to be the frontiers of bottom-up participatory design. This review will be followed by two antecedent projects: Medical Faculty Housing by Lucien Kroll (1976) and Cedric Price's unrealized Fun Place (1965) through which a wide range of participation forms were conceived. In both of these projects, the designers attempted to use various industrial and computational methods for augmenting participatory design processes. However, the capacity and potentials of these were limited at the time.

Acknowledging this gap, this paper aims to discuss the potentials of modes of participatory design in relation to the latest developments in Information and Communication Technologies (ICT). Regarding this aim, the research questions to be explored in this paper are:

1. What are the participatory design strategies used in the prominent bottom-up cases?
2. Which forms of ICT use were conceived in these?
3. Can these serve as a basis for a novel approach that incorporates state-of-the-art ICT?

In this context, the first part of the study (Section 2) involves the extraction of basic strategies and principles from the antecedents of bottom-up participation. Building on the findings of this case study, the second part (Section 3), reframes the existing ICT tools and strategies in terms of space and time and elaborates on their potentials for supporting the revealed modes of participation. The following section introduces a framework for ICT-enabled design participation which exploits a set of digital and non-digital collective action tools to support sustainable ways of self-organization and bottom-up design. As a result (Section 4), by distilling these findings, the last part of the study reveals the future potentials of ICT-enabled bottom-up participation including novel methods such as crowdsourcing, crowdfunding and emergent technologies such as smart structures and fabrication.

BACKGROUND: LEARNING FROM THE PAST
As briefly introduced above, it is possible to identify two main strands of city-making from the perspective of governance. Top-down approaches are based on the conception of citizen participation as a state-centered practice. In contrast, bottom-up approaches are characterized by social activism and civic engagement in the absence of higher level direction (Bowns and da Silva, 2011). Although these models seem to be conflicting with each other, bottom-up and top-down practices are often used in combination during the production processes of the cities. It is impossible to find a practice which purely fits in one category (Salingaros, 2005, p. 223).

After the beginning of the twentieth century, with the introduction of the modernist approaches, authorities in the Western World increasingly adopted strict predefined territorial boundaries and zones, and predominantly top-down planning methods which increasingly disabled the community participation (Nour, 2015, p.21). At the beginning of the 1960s, architecture was in a severe crisis. The problems of modernist functionalism and bland architecture emerged as a consequence of economically-driven utilitarianism and these became more and more evident in the media (Blundell-Jones, 2005). A monumental example was the Pruitt-Igoe Housing located in St. Louis, Missouri, USA, in which the living standards started to decline only after two years and eventually, in less than eighteen years, it had to be demolished due to its social and physical unsustainability (Larsen and Kirkendal, 2004).

The decades that followed these developments gave birth to new approaches criticizing modernist top-down practices. Through specific design cases, the architects aimed at the bottom-up involvement of users in alternative ways and made extraordinary experiments which served as a model or stimulus for later developments. In this section, I will make an attempt to relate these cases to each other and extract the common strategies behind these.

Frontiers of Bottom-up: Experimental Practices after the 1960s
Habraken (1961) was one of the first to introduce the idea of spatial self-determination as a citizen right. In his book "Supports: an alternative to mass housing" he suggested a participatory design method based on the decoupling of the support infrastructure and the "infill". In this method, while the infrastructure is framed as a static and long-term investment, individual habitation units were envisioned to be customized by the users with the help of an architect.

Habraken's concept of a supporting/enabling megastructure was not novel at that time (Figure 2); it can be traced back to Nieuwenhuys' New Babylon experiments between 1956 and 1978, Friedman's Mobile Architecture in 1958 and Fuller's structural designs in 1955 (Lobsinger,

2000). His ideas also resonated with Jane Jacob's (1961) coeval critique of planning practices in the USA and the Situationists (Mathews, 2006).

Yona Friedman, *Spatial City*, 1959

Constant Nieuwenhuys, *New Babylon*, 1959-1970

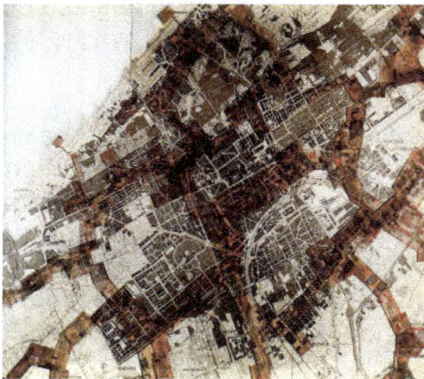

Buckminster Fuller, *Airplane Hangar,* 1955

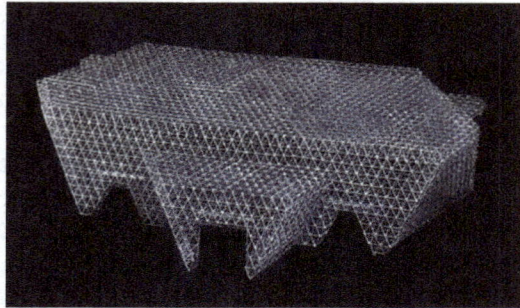

Figure 2. The exploration of the idea of supporting/enabling megastructure in various experimental practices after the Second World War (Source: Author).

The real contribution of Habraken was his ability to combine the criticism of modern architecture and bottom-up participatory practices with an alternative industrial vision. The methodological studies of Habraken's office SAR (Stichting Architecten Research) that followed the lines set in his book increased his impact on the world of architecture and inspired several architects including Lucien Kroll. Even today, this school of thought is still active under the name of "open building network" in close collaboration with the building industry (Kendall, 2015).

The most classic example following a similar design approach as Habraken's supporting/enabling structure was Le Corbusier's Unité d'habitation, in Firminy France (1962). This design employed a "bottle rack" principle: an open structural frame infilled with different housing types (Schneider and Till, 2007, p.168). However, the central focus of this project was not necessarily on bottom-up design but rather on creating variety and opening up possibilities for choice

In Poland, Oskar Hansen, a member of the group of architects Team X, was one of the first critical voices regarding the orthodoxy of modern Athens Charter and the followers of Le Corbusier. He presented his "open theory" to the founding meeting of the group Otterlo in 1959 (MACBA, 2015). This "attitude" (rather than a theory) was conceived initially as a tool for the design of architectural projects, although evolution and its application in the fields of education, editing films, the games and visual performance practice led to broad set experiments that were interacting with each other, sharing and socializing through art objects. Hansen experimented with strategies open to uncertainty, flexibility and collective participation. He coined the term "open form" to describe architecture open to the possibility of continuous transformation, open to the influence of nearby practices as well as to the approval of the users (MACBA, 2015). One of the most memorable products of Hansen was the design of adaptable furniture which allows almost infinite uses of a rectangular living room.

In parallel with these contributions, in 1962, Walter Segal developed a low-cost housing solution suitable for self-build, while trying to address his own problem of providing a temporary home for his family. This practice evolved into the development of "The Segal Method" and several participatory design projects in which 27 families worked with architects to design and build their own homes (Broome, 1995).

One of the most interesting bottom-up cases in the same era was the building of the Gladsaxe playground in 1969 (Gehl and Svarre, 2013). Led by Jan Gehl's team, residents from Høje Gladsaxe, a newly built public housing complex in a suburb of Copenhagen, Denmark and the students from Copenhagen universities ventured into an unauthorized construction of a large playground on an empty stretch of gravel in front of the multi-story complex (Figure 3). According to Gehl and Svarre (2013) the playground was perceived as quite successful while it was being built and for many years later. This case was recorded as one of the earliest examples of truly bottom-up design and construction of a public space in the Western world, triggered by an architect/ urban designer.

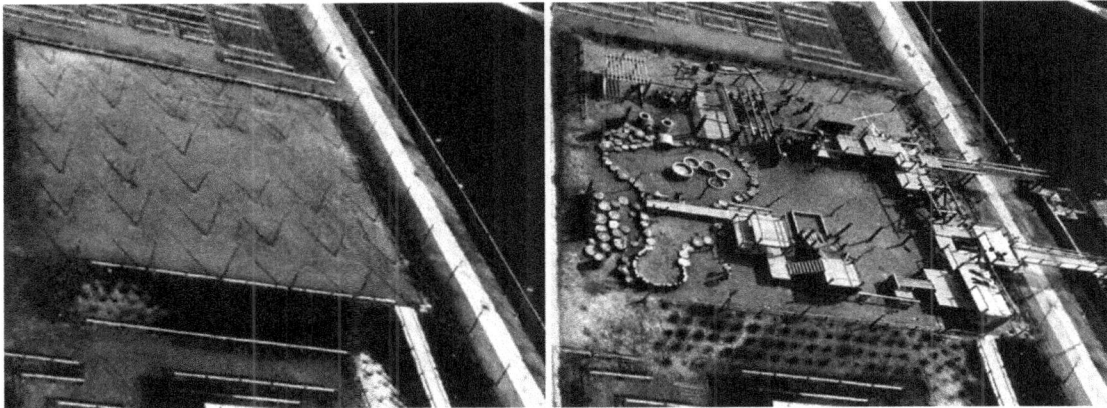

Figure 3. Høje Gladsaxe Playground (1969) built in only one day by the residents and university students (Gehl and Svarre, 2013).

Around the same time period (1969-1978), Lucien Kroll orchestrated the design of the Medical Faculty of the Catholic University of Louvain (UCL) with the student organization "La Maison Medicale-La MéMé". In the following text, I will make an in-depth review of this case.

LA MÉMÉ (IN-DEPTH CASE 1)

The project was initiated when the Catholic University of Louvain (UCL) decided to move its Medical Faculty to Brussels, Saint-Lambrechts-Woluwe. The university authorities made an exceptional decision and presented the preliminary design of the Medical Faculty Housing to the student committee. The students rejected the project and contacted Kroll for his services (Kroll, 1987). As a close follower of Pierre Bourdieu, Kroll took the task and questioned every aspect of the institutionalized practices with the contributions of the spirited students of UCL. He intended to create an open design process, "an action open to new necessities and to decisions that are always provisional and incomplete" (Kroll, 1987). He aimed at establishing an intellectual climate through which a kind of friendly organization would emerge to result in a homeopathic kind of architecture (Kroll, 2005).

Kroll organized meetings with the committees and discussion groups. In these meetings, he received conflicting ideas. Instead of flattening out all the differences of approach and attitudes he tried to incorporate them into the design process (Kroll, 1997). This was a creative refutation of the idea of "consensus". Throughout the project, the students were empowered to participate in two forms: through getting involved in the design process and through the participation opportunities provided by the architectural design *per se* (Figure 4).

Figure 4. Medical Faculty Student Housing by Lucien Kroll. The users
participated in the decision process, and the architectural product enabled
them to shape and reshape their surroundings (Source: Author).

Kroll developed a flexible structure system which he called "wandering columns" based on a loosely defined grid. He collaborated with a professor of computer engineering to manipulate the grid to support the irregular and heterogeneous shell of the building. He designed the artificial ground around the project to provide raw space for further development (the aspects of wandering columns and his long-term vision for expansion are more evident in the Alma Metro Station, which was built as an extension of the project).

The "infill" –inherited from Habraken– is hypothetically removable: demountable window frames, moveable partitions, and prefabricated sanitary units. The architect used his own interpretation of the Habraken's SAR module but refuted the idea of functional zones (Kroll, 1987). According to the principles of co-habitation, the infill can be torn down by the users, which encourages them to take initiative in planning and re-planning their environments. The plan would always be incomplete.

In La MéMé, Kroll did not see aesthetics as the central point of design. Through this project, he strongly criticized what he called the "easel architecture": aesthetically pleasing but isolated from the people, culture, and community. In his book "Architecture of Complexity", Kroll (1987) reserved a whole chapter to the computers. Instead of computer-aided design (CAD), he suggested computer use in design (CUD) as a more appropriate term for describing his vision. He stressed the importance of open-endedness and heavily criticized the inflexible artificial intelligence practices of the time that led to self-contained, closed and repetitive results.

In contrast, Kroll envisioned the drafting software as a potential tool that allows open-endedness through which the architectural product and the social relationships can be involved in the design and manufacturing process. However, the communication technologies were not developed enough to realize fully the social part of the potential.

The computer-based social interactions he foresaw were limited to three-dimensional drawings which he found useful for the communication of early ideas to the inhabitants. He suggested that infinite interactions were required to deal with the infinite diversity of the real world.

In close contact with the users, he employed various algorithms to create diversity and differentiation and presented a library of components that can be combined according to user needs (Kroll, 1987). He tested a computational method –anthropomorphism– to allow a type of architecture with variant building programs and devised the role of the architect as a developer of "types" which can be varied by the inhabitant. The diversity of the outcomes was unmanageable due to the technological limits of the time. As a result, he worried that the process would lead to the Taylorist practices that he criticized (Kroll, 2013).

In the same chapter relating to the ICT use, Kroll (1987) described another possible role for the computer: evaluation and modification. During the design process, a custom program provided comparisons between the choices of components designed by the architect and enabled rapid updates of these particular components throughout the process. Furthermore, by creating associated representations of the components, Kroll used a computer to generate façade drawings to be revised and detailed by the designer. However, he stressed that this process can never be reliant on automation, which Kroll found "an absurd and unhealthy claim".

In conclusion, Kroll created an ambitious piece of "anarchitecture", challenging every possible aspect of the architectural practices of the time. It became an "icon of democratic architecture" (Poletti, 2010) for Kroll's alteration of the usual hierarchical relationship between the architect and the user during the process –and most importantly– the development of novel design interventions to enable bottom-up participation. As Jencks (2011) suggests, although his ideas were not realized to the anticipated extent (both in terms of the process and the product) Kroll's importance in participatory design history can never be exaggerated. For some critics, La MéMé was the absolute denial of architecture (Kroll, 1997). However, Kroll was not the only architect who made a significant effect on the future bottom-up practices.

THE FUN PALACE (IN-DEPTH CASE 2)

In the intellectual climate of the 1960s described above, director Joan Littlewood commissioned Cedric Price to design an informal and dynamic entertainment center: the Fun Palace. It was conceived to be permanently under construction meant to empower the ordinary citizens to be active participants in a never-ending and reflexive play (Banham et. al. 1969). The extraordinary nature of the project came from the wide range of interdisciplinary contributions of Gordon Pask (cybernetics), Buckminster Fuller (structural design), Yehudi Menuhin (symphonic music) and Reyner Banham (architectural theory) (Lobsinger, 2000).

The project aimed at fusing information and communication technologies and industrial building principles "to produce a machine capable of adapting to the needs of users" (Price, 1965). In contrast with Kroll's refutation of Le Corbusier's metaphor of architecture as a "machine-to-live-in", Cedric Price adopted and developed it further. The project was an attempt at exploring improvisational architecture with the means of cybernetics and information technologies (Mathews, 2005).

Fun Palace did not have a fixed floor plan and intended to "encourage random movement and variable activities" (Lobsinger, 2000). Mobile components such as flying escalators, walkways, and activity enclosures were carried by a megastructure and transported by a crane when necessary (CCA, 2015). The suggested time and place specific facilities covered jam sessions, dance and science playgrounds, teaching film, drama therapy, modeling and making areas and music stations with instruments on loan (Landau, 1984). Similar to La MéMé, Fun Palace was not primarily an aesthetical exploration. The building was conceived to be super-functional and adapt to the people's needs in a sustainable manner (Figure 5).

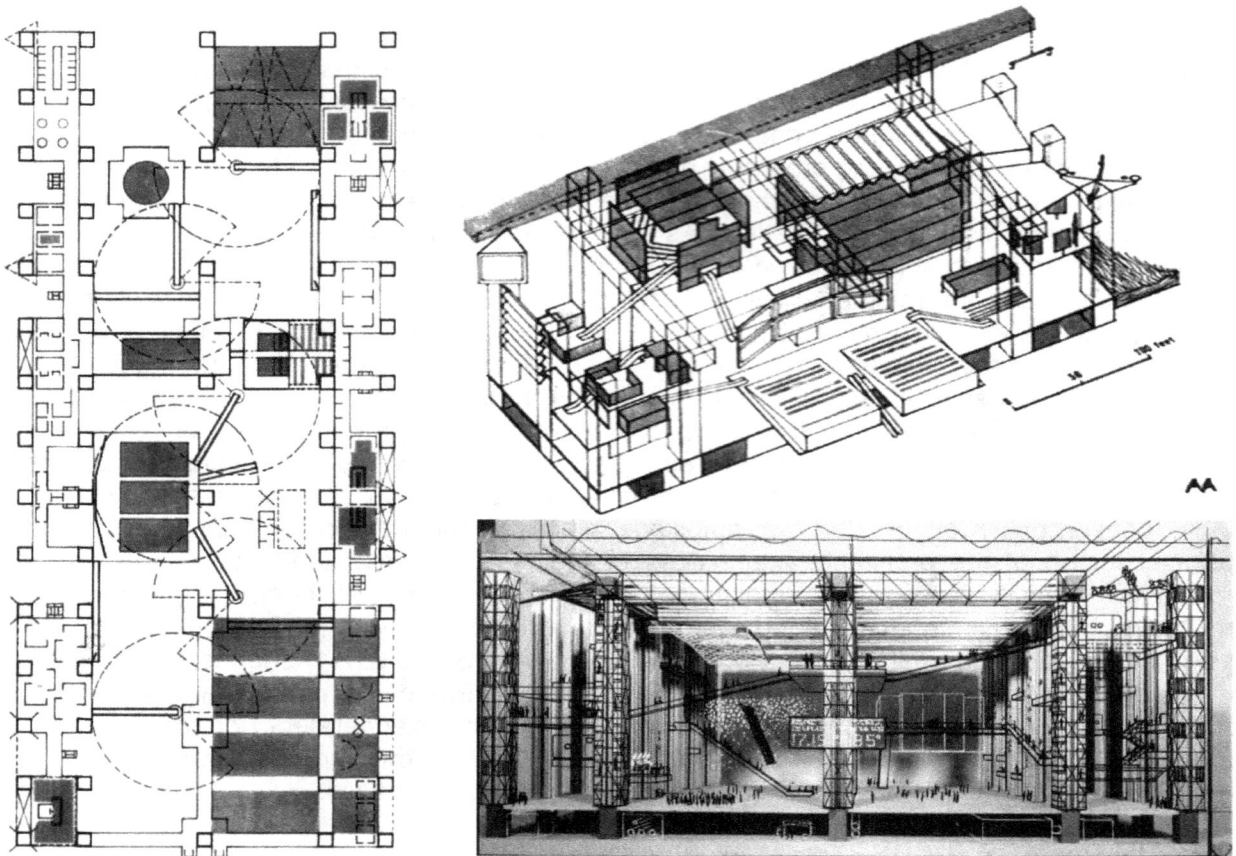

Figure 5. The Fun Palace project by Cedric Price and Joan Littlewood proposed a dynamic program that joins ICT and industrial building principles to produce architecture capable of adapting to the needs of the users. Illustrations: CCA Library Database (2015).

As introduced in the previous section, the interdisciplinary design team which Price collaborated with included an English cybernetician and psychologist, Gordon Pask. During the project, several practices were proposed by Pask for the cybernetic regulation of day-to-day activities (Mathews, 2006). In this sense, Fun Palace would be an ongoing conversation between the building and its users – "an assemblage of interactive systems of interaction" (Harding, 2008).

Pask (1969) defined a number of domains of interest for cybernetic interventions. Among those were the Fun Palace and environment, visiting patterns, mechanical and architectural considerations, provision of specific participant activities, interactive activities, individual participant situations (teaching machines), controlled group activities, conditioning systems and cybernetic art forms (Mathews, 2005). As a proof of concept, Pask created an apparatus to collect feedback from the users after the realization of the project. The proposed tool was a physical communication system which he planned to be used informally in one of the theaters to "accommodate an invited audience" (Pask, 1969). The audience would be responding to a variety of activities using this tool and would be able to transform the theater based on their preferences. Through this exercise, Pask questioned to role of the users and explored novel ways of participation in an open-ended and performative manner.

The Fun Palace was never realized, but it is still known as one of the most prominent participatory design cases to inspire numerous architects including Richard Rogers' and Renzo Piano's Pompidou Center (1976) which closely resembles the initial sketches of Price.

In conclusion, Kroll's and Price's works can be considered as *prototypes* of participatory, bottom-up architectural design. However, it is necessary to differentiate between these two projects. First of all, Price and his team failed to realize the Fun Palace. Although it was designed to be built, it can be seen as *a proof of concept* for a utopic project. On the other hand, La MéMé was partially realized and served as a *semi-functional prototype* through which many inspiring ideas were experimented. It still stands in Brussels as a *heterotopia* between the ideal and the real, frozen in time.

DISCUSSION: MODES AND STRATEGIES FOR BOTTOM-UP PARTICIPATORY DESIGN

Reflecting on the typical characteristics of vernacular architecture reviewed in Section 1, it is possible to claim that there are significant similarities between the basic principles employed in vernacular architecture and bottom-up participatory design practices after the 1960s. Among these, the most recurrent ones are *flexibility, adaptability, and self-organization*. However, the nature of these practices is quite different.

The vernacular cases illustrate ventures of collective bottom-up activity leading to an anonymous construction. On the other hand, in the latter cases architects play a central role as a "co-designer-enabler". From this perspective, in these cases architects aimed to combine top-down and bottom-up activities and facilitate a bottom-up participatory design process. In parallel, they intended to empower users through the design itself through some form of consultancy. In this sense, it is possible to derive two interconnected modes of participation from the reviewed cases:

1. Participation {in} the design process
2. Participation {through} the design product

These two modes were significant in the ways they enable the users and architects to co-produce architectural designs in a sustainable and participatory manner. To start with, Participation {in} the design process is similar to today's widely recognized interpretation. It involves practices that "allow various actors to contribute to the overlapping phases of the planning and decision–making" (Horelli & Wallin, 2010).

In the case of La MéMé, Kroll has arranged numerous meetings with committees and discussion groups to empower the student groups (although the level of participation and openness were challenged in the following years). Price, on the other hand, did not believe that the user needs can be precisely forecasted. The user participation model he conceived would take place post-occupancy. However, he shared his authority with several intellectuals such as Littlewood, who acted as an essential part of the design team. Instead of pursuing traditional consultation meetings, he asked for the participation of an interdisciplinary committee to collaboratively design an enabling type of architecture that facilitates participation to the greatest known extent.

In this context, action research orchestrated by Jan Gehl in the Høje Gladsaxe Playground was one of the extreme cases. The design emerged directly from the user needs and built by the residents in the area. It was a constructive revolt on top-down planning approaches which created a long-term impact on public space. The second and the most interesting mode observed in the presented cases is participation {through} the design product. This kind of empowerment takes place when various spatial qualities of the architecture enable the inhabitants to shape and reshape their own living environments. As reviewed in the previous section, several terms were used to describe this kind of participatory approaches, among those were: "open form", "open design process" or "open building".

In both of the in-depth cases presented above, Kroll and Price aimed at the participatory creation of infinitely flexible interactive spaces which represent the diversity of the needs of the inhabitants. The forms of their designs were intended to be altered to accommodate the changing needs of the users. In the La MéMé case, the dynamic elements were the "infil": demountable window frames, moveable partitions, and prefabricated sanitary units. The Fun Palace project envisioned mobile components such as flying stairs, walkways, and modular activity enclosures.

Furthermore, in both of the cases, a structure independent from the infill was used for facilitating the dynamism of the architectural program. Besides the participatory modes discussed above, it is possible to identify several strategies for recurrent in the bottom-up practices. Among those the most prominent ones are:

- Orchestrated self-organization
- Intense focus on the impact which architecture can make on the users
- Incorporation of user variety and differences into the architectural design process and the product
- Incomplete, dynamic program as an enabler for the continuous representation of the user needs
- Embracing spontaneity and improvisation in the design process
- Development of design rules or systems that "regulate" the building in an open-ended way
- Long-term vision: flexibility, adaptability, and polyvalence
- Reflexivity in terms of viewing the everyday life as a site for transformative spatial practice

In the next section, I will discuss how these can be strategies which can be employed in an effective manner using cutting-edge ICT tools and methods.

FUTURE POTENTIALS OF ICT-ENABLED BOTTOM-UP PARTICIPATION

Building on the strategies and the participation modes introduced above, participation van be understood as a reflective self-organization practice which includes interactions {in} the design process as well as {through} the design product. In this approach, ICT tools fuse two cycles of cooperation in which the output of one process is transformed by a second process and transferred to the other one as input. In contrast with completely digitalized mode of operation this suggests the augmentation and enhancement of traditional participatory practices through the use of ICT tools and strategies.

The first cycle involves a type of social knowledge construction, building social capital. This capital is transferred to the second cycle through which the users gather resources, take action, accumulate experience and give feedback to the first cycle. This open-ended process involves several techniques which can be supported by different types of novel ICT-enabled participation accessible today:

- Crowdsourcing
- Crowdfunding
- Responsive structures
- Fabrication and low-cost manufacturing

Each participation technique is appropriate for specific levels of civic engagement (Megahed, 2014, p.104). In the following part of this paper, I will try to explain the relevance of the above with solid examples.

Crowdsourcing: Collecting Feedback, Ideas, and Information from the Users

The last decade has witnessed the proliferation of new web-based social software and information aggregation services which facilitate social knowledge construction. These are commonly put under the umbrella of the term "Web 2.0" which has been described in the manifesto of O'Reilly issued in 2005 as *"practices in which web is used as a platform for harnessing collective intelligence, delivered as a service, not a product, based on lightweight programming models, backed by a specialized database, supporting PC and non-PC devices and providing a rich user experience"* (O'Reilly, 2005). Relying on a combination of web 2.0-based social software and information aggregation services, Geoweb 2.0 technologies stand as a strong alternative to the traditional linear and hierarchical knowledge production methods. They are loaded with constructivist learning, and production principles applied both in the making of the facilitating open-source environments, and the ways they enable social knowledge construction. In this sense, they are well positioned to act as a medium for facilitating dialogue and learning as well as bottom-up communicative action. The real power of Geoweb 2.0 comes from the way it is utilized for the inclusion of knowledge acquired through lived experience or experiential knowledge; which had been granted less legitimacy in the past (Elwood, 2006).

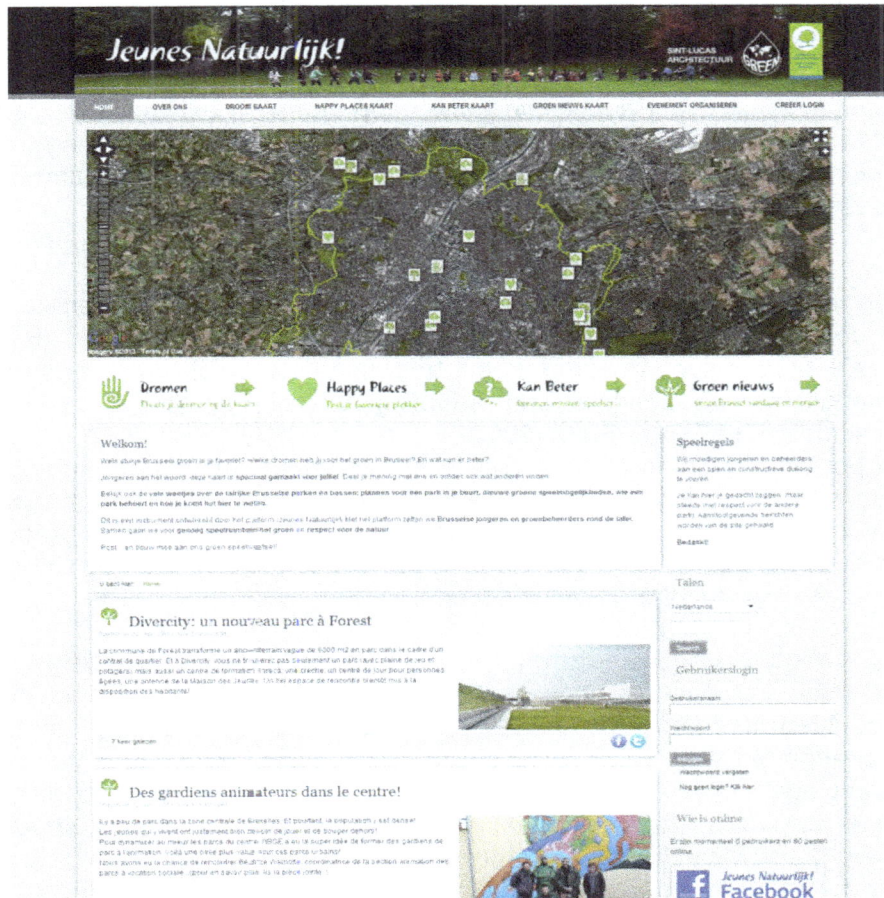

Figure 6. An example of a crowdsourcing interface for collecting user ideas, preferences, and socio-spatial problems from the perspective of the users (Pak and Verbeke, 2014).

The democratic promise of crowdsourcing practices is that complex problems can be adequately addressed by harnessing the wisdom of the crowds, such as by allowing the general public to formulate their needs, problems, opinions (Figure 6), and even solutions themselves (Surowiecki, 2004).

Crowdfunding: Gathering Resources and Endorsement

Crowdfunding is an emerging method which provides novel ways to empower users and designers over the internet to obtain funding for the projects they want to endorse. During the recent years several community-based web applications have successfully managed to accomplish this goal. For instance, according to the Architizer website (2015), since 2009, Kickstarter Crowdfunding platform has collected more than $660 million to support various types of projects, ranging from film productions to food industry to the design and development of technology and devices. In this sense crowdfunding had significant potentials for activating bottom-up change, specifically the *resources* and *intervention* steps referenced in Figure 6.

In order to gather resources in a bottom-up manner, at least four types of crowdfunding are identified (UK Crowdfunding Association, 2015):

- Donation crowdfunding: The users invest because they believe in the cause and donate without the expectance of anything of tangible value in return.
- Reward crowdfunding: In this model simple rewards are offered such as early access to the endorsed products.
- Debt Crowdfunding: Investors receive their money back with interest. Also called peer-to-peer (p2p) lending.
- Equity crowdfunding: People invest in an opportunity in exchange for equity.

These provide a wide space for action for the bottom-up practices, ranging from communal living and joint ownership ventures to the use of endorsements as an indication commitment to the suggested ideas.

Responsive Structures: Dynamic Interventions adapting to user needs

Reflecting on the latest developments in ICT, it is possible to claim that sensor networks and smart structures can play an important role in the gathering of feedback as well as the support of user interventions. Responsive structures is an emerging field which involves measuring actual environmental conditions via sensors and adapting their form, shape, color or character responsively via actuators (d'Estrée Sterk, 2009).

The biggest potential of these technologies is the establishment of an ongoing conversation between the building and its users as described by Pask (1969) as "an assemblage of interactive systems of interaction". In this sense such structures can afford to encourage random movement and variable activities as well as time and place specific facilities foreseen by Cedric Price in the Fun Palace Project reviewed in the previous section.

Fabrication and low-cost robotic manufacturing: enabling user interventions

Low-cost robotic manufacturing methods have a potential to unlock self-production practices, which can also be integrated into the proposed model after the crowdfunding step. In this sense, 3D printing can change the way we produce buildings and building components. These methods involve the assemblage of units by depositing thin layers of material such as plastic, metal, concrete and even ceramics, therefore making it possible to build product structures that are strong yet lightweight which can be carried around by the users.

In this context, the ability to produce and reproduce the "infill" can empower the users to shape and reshape their living environments. The dream of a dynamic plan would be possible as a result of this evolution.

However, at the moment, these technologies are far from being affordable. Different low-cost techniques are still under development. Moreover, existing regulations also poses a challenge to these practices since they are based on a static understanding of space. Technologies such as radio-frequency identification (RFID) can help to keep a dynamic representation of building elements. This will be possible through the wireless use of electromagnetic fields to transfer data and automatic location and identification of tags attached to objects. These can be components such as walls and windows, as well as furniture. These will enable a new way of designing through "a plan that draws itself" as the users continuously reshape their living environments.

A Hypothetical Use Case in Practice

This part of the paper takes a large-scale housing project as a hypothetical case for demonstrating the potential of the ICT-enabled participation techniques presented above.

Following the cooperation cycles introduced in the previous section, the participatory design process starts with *crowdsourcing* through which the needs and requirements are collected in a structured manner. Then, these are converted into several alternative design *ideas* and *integrated* into the context by the architect, with the continuous ICT-enabled feedback of the users.

Afterwards the users are asked to fund the project through *crowdfunding*. If the process succeeds, the project gets transferred to the second cycle and constructed with the contribution of the users. If the funding process fails, the design process cycle is repeated: through a new crowdsourcing practice, the user feedback is collected to identify the problems of the design, and to develop new alternative projects, which will then be asked to be *crowdfunded* by the users.

The contribution of the model becomes more evident *after* the construction of the architectural project. Following the experiences of the residents, *post-occupancy feedback* is collected through crowdsourcing.

When necessary, novel ideations on how to improve the architectural design are created with the continuous ICT-enabled feedback of the users and integrated into the existing context. Examples of these ideas can be making interventions regarding the communal or personal spaces, removing/adding partitions or reconfiguring the rules for co-habitation.

Afterwards, the users are again asked to fund the changes through crowdfunding. If there is enough support, the process moves to the second cycle and suggested interventions are made. Following the intervention, the user feedback is collected, and the participation continues to take place when necessary. In this context, it becomes possible to develop habitats which can adapt to the users' needs in a sustainable manner.

Reflections on the Future of Architectural Design Education

While the strategies and tools presented above have the potential to empower the citizens in design practices, they can also serve to facilitate and augment novel educational approaches in design education. For instance, a **crowdsourcing** platform can be utilized as an infrastructure to implement the "community-based design learning model" introduced by Salama (2015, p.116). The first possibility to accomplish this goal is to position the referenced platform as an interface enabling learning from the inhabitants, non-governmental organizations and local experts (Pak and Verbeke, 2012). Such practices can enable learning through various research tracks:

- Understanding the socio-spatial complexity through online participatory mapping
- Learning about the spatial issues relevant to a specific area by a geo-located analysis on how people manifest their identities and appropriate spaces
- Creating simple challenges to motivate users to respond and act; and identification of their needs through this practice (Salama, 2015, p. 171)
- Stimulation of rigorous research strategies combining personal observations with participatory data and interviews (triangulation) (Loopmans et al., 2011)

Specifically, in urban contexts such as Brussels, London and Istanbul super-diversity brings many challenges to spatial design and creates conflicts due to the overlapping needs (Vertovec, 2007). In this sense, crowdsourcing can specifically be fruitful for students to address this complexity and help them in the design process.

A second possibility is to position crowdsourcing platforms as an interface for the mediation of the dialogue between the design students and studio teachers. Particularly in design studios with a large number (50+) of students, the platform can be configured to support, augment and enrich the reflective learning processes (Pak and Verbeke, 2014) by enabling students and teachers to provide feedback to each other to improve their performance (Salama and El-Attar, 2010).

Furthermore, following crowdsourcing, through **crowdfunding,** design ideas produced by the students can be opened up to public for endorsement and developed further through the cycles introduced in the previous sections. In the case of relatively small scale interventions (e.g. urban furniture, simple artistic interventions, parklets]) raised funds can be sufficient enough to realize these ideas. For larger scale projects (e.g. public squares and community centers), public endorsement can be interpreted as support from the public and used as leverage to promote further the projects to the authorities.

Finally, low-cost fabrication tools such as laser cutters can enable the students to make prototypes and test in a real-world context, gather feedback from the users and improve their designs (Nys, 2015). Specifically, in disadvantaged urban areas and the developing world these kinds of practices can provide the students and local citizens various opportunities to participate in the shaping of the urban environment with the contributions of the civil society as well as other relevant actors.

ACKNOWLEDGEMENTS

This research partially includes findings from a three-year type (B) "Prospective Research for Brussels" postdoctoral research grant from the Brussels Capital Regional Government, the Brussels Institute for Research and Innovation (INNOVIRIS) awarded to Burak Pak, promoted by Johan Verbeke. The author would like to thank Lucien Kroll and Dag Boutsen for their valuable feedback during the preparation of this paper.

REFERENCES

Ashby, W.R. (1958). Requisite Variety and its implications for the control of complex systems, *Cybernetica* (Namur) Vo1 1, No 2.

Bauman, Z. (2000). Liquid Modernity, Polity Press, Cambridge.

Broome, J. (1995). Segal Method Revisited, Architects Journal 202, no. 20,

Banham R., Barker P., Hall, P. Price, C. (1969). Non-Plan: an experiment in freedom', *New Society*, 13 (338).

Belluschi, P. (2012). The Meaning of Regionalism in Architecture, in Canizaro, Vincent B. ed. Architectural Regionalism: Collected Writings on Place, Identity, Modernity and Tradition. New York: Princeton Architectural Press.

Bowns, C., da Silva, C. P. C (2011). Community Practice, the Millennium Development Goals and Civil Society Measures in Brazil, *Archnet-IJAR, International Journal of Architectural Research*, 5(2), pp. 07-23.

CCA Library Database (2015). Fun Palace (Project), Accessed on 01/05/2015 http://cel.cca.qc.ca/.

d'Estrée Sterk, T. (2009). Introduction: Thoughts for Gen X-Speculating about the Rise of Continuous Measurement in Architecture, ACADIA 09: reForm( ) - Building a Better Tomorrow [Proceedings of the 29th Annual Conference of the Association for Computer-Aided Design in Architecture (ACADIA) Chicago (Illinois).

Elwood, S. (2006). Negotiating knowledge production: The everyday inclusions, exclusions, and contradictions of participatory GIS research, *The Professional Geographer*, 58(2), pp. 197-208.

Foerster, H. (1974). Cybernetics of Cybernetics, Urbana Illinois: University of Illinois.

Gehl, J. and Svarre, B. (2013). How to Study Public Life, Island Press.

Habraken, J. N. (1961). De dragers en de mensen: Het einde van de massawoningbouw, Scheltema & Holkema.

Harding, J. (2008). Architectural Systems, An Essay based on the initial text "Cybernetics and the Mangle" by Andrew Pickering. Architecture, Computing & Design, CECA, Accessed on 05/05/2015 http://people.bath.ac.uk/jeh33/Publications/Harding.

Horelli, L.. Wallin, S. (2010). The Future-Making Assessment Approach as a Tool for E-Planning and Community Development: the Case of Ubiquitous Helsinki. In Silva, C. N. (Ed.), Handbook of Research on E-Planning: ICTs for Urban Development and Monitoring (pp.58-79). Hershey, PA: IGI Global.

Jencks, C. (2011). The Story of Post-Modernism: Five Decades of the Ironic, Iconic and Critical in Architecture. London: Wiley.

Kroll, L. (1987). The Architecture of Complexity, P. Blundell-Jones MIT Press.

Kroll, L. (1997). Anarchitektuur, in Componenten 2, Omtrent de Modernisering van de Architektuur, J.D. Besch (ed) Publikatiebureau Bouwkunde Delft.

Kroll, L. (2005). Animal town planning and homeopathic architecture in Architecture and Participation Blundell J. (ed.) Taylor and Francis. Kindle Edition.

Kroll, L. (2013). Friendly Architecture in OASE Journal No. 90 Nai Publishing, Rotterdam.

Landau, R. (1984). A Philosophy of Enabling: Cedric Price, Works II. Cedric Price. London: Architectural Association, 9-1.

Larsen, L. H., Kirkendall, R.S. (2004). A History of Missouri: 1953 to 2003. University of Missouri Press. ISBN 978-0-8262-1546-8.

Lobsinger, M. L. (2000). Cybernetic Theory and the Architecture of the Performance: Cedric Price's Fun Palace," in Anxious Modernism, Goldhagan J and Legault R., eds., Cambridge, MA: MIT Press.

Loopmans, M., Leclercq, E., Newton, C., Raeymaekers, K. (2011). Plannen voor mensen : handbook sociaalruimtelijke planning, Antwerpen, Garant.

Mathews, S. (2006). The Fun Palace as Virtual Architecture Cedric Price and the Practices of Indeterminacy, *Journal of Architectural Education*, 59(3).

MACBA (2015). Oskar Hansen, Forma Oberta, Exposició organitzada I produïda per: Museu d'Art Contemporani de Barcelona (MACBA) i Museu d'Art Modern de Varsòvia accessed on 19/04/2015 http://www.macba.cat/uploads/20140709/CAT_NdP_Oskar_Hansen.pdf

Megahed, N. (2014). Heritage-Based Sustainability in Port Said: Classification of Styles and Future Development, *Archnet-IJAR, International Journal of Architectural Research* 8(1) pp.94-107.

Nour, H. (2015). RECONSIDERING THE WAQF: Traditional Mechanism of Urban Regeneration in Historic Muslim Cities, *Archnet-IJAR, International Journal of Architectural Research* 9(1) pp.18-30.

Nys, J. (2015). Reframing Fragile Heritage – Micro-strategies for a social and sustainable re-use of fragile heritage. Ph.D. Thesis under preparation, KU Leuven Faculty of Architecture.

O'Reilly (2005). What is Web 2.0, accessed on 19/04/2015 http://oreilly.com/pub/a/web2/archive/what-is-web-20.html.

Pak, B., Verbeke, J. (2012). Design Studio 2.0: Augmenting Reflective Architectural Design Learning. Electronic Journal of Information Technology in Construction, 17 (Special Issue eLearning 2.0: Web 2.0-based social learning in built environment), 502-519.

Pak, B., Verbeke, J. (2014). Geoweb 2.0 for Participatory Urban Design: Affordances and Critical Success Factors. International Journal of Architectural Computing, 12(3), 283-305.

Pask, G. (1969). The architectural relevance of cybernetics. *Architectural Design* 9.

Poletti, R. (2010). Lucien Kroll:Utopia interrupted, *Domus* 937, June 2010.

Price C. (1965). The Fun Palace Project, *Architectural Review* 74.

Rapoport, A. (1969). House Form and Culture (Foundations of Cultural Geography Series). Englewood Cliffs, N.J.: Prentice-Hall.

Rudofsky, B. (1964). Architecture without Architects, New York: Museum of Modern Art.

Salama, A.M., El-Attar, S.T. (2010) Student Perceptions of the Architectural Design Jury, *Archnet-IJAR, International Journal of Architectural Research*, 4(2-3) pp.174-200.

Salama A. M. (2015). *Spatial Design Education: New Directions for Pedagogy in Architecture and Beyond*, Farnham, Surrey: Ashgate Publishing Limited.

Salingaros, N. A. (2005). Principles of Urban Structure, Delft: Techne and Delft TU.

UK Crowdfunding Association, (2015). UKFCA / What is Crowdfunding, accessed on 05/05/2015 http://www.ukcfa.org.uk/

Vertovec, S. (2007). The emergence of super-diversity in Britain, Centre on Migration, Policy and Society, Oxford University Working paper no. 25.

AUTHOR
Dr. Arch. Burak Pak
Post-doctoral Research Fellow
University of Leuven, Faculty of Architecture, Brussels, Belgium
Email address: burak.pak@kuleuven.be

4

ARCHAEOLOGY, ARCHITECTURE AND CITY:
THE ENHANCEMENT PROJECT OF THE ARCHAEOLOGICAL PARK OF
THE BATHS OF BAIAE

Renato Capozzi*, **Adelina Picone****, **Federica Visconti*****
DiARC_Department of Architecture,
University of Naples "Federico II", *Italy*

*renato.capozzi@unina.it
**adelina.picone@unina.it
***federica.visconti@unina.it

Abstract
Following the theoretical and disciplinary framing of the elements that substantiate the relationship of archaeology with architecture and the city in light of the transformations of the modern city, the project aims at valorizing the archaeological asset, promoting a knowledge of the ruins from multiple theoretical perspectives. The enhancement project of the Archaeological Park of Baiae experiments with different modalities of knowing that include the knowledge of the relationship between the ruin and the landscape, the philological, typological-constructive knowledge, and the knowledge of the ruin's own spatial elements. Bringing together the contributions of different disciplines and experts under the coordination of an architect, the theoretical core of the project promotes the enhancement of the Archaeological Park, envisioning it as a means of valorisation of a wider urban environment.

Keywords: *Archaeology; Architecture; City; Interdisciplinarity*

INTRODUCTION: THEORETICAL FRAMEWORK

The contemporary city with its architecture, its multiple infrastructural systems, its yearning to change, its challenge to welcome climatic and environmental change, finds its meaning in the relationship with history and memory, with archaeology and the monuments. The idea that the monuments of the contemporary city, whether archaeological remains or permanent elements of the settlement structure, should, as sites of memory, function to inspire development and change is a shared and acknowledged notion. The codification and evolution of "urban archaeology" (Francovich, Manacorda, 2000) into a discipline and the large bibliography available on this topic mobilize scientific methodologies to govern the processes. The definition of urban archaeology - «The archaeological research in an existing city over its whole settling history from the foundation to the present without privileging a period over another» - refers to a necessarily systematized knowledge of the city's archaeological history that includes excavations and interdisciplinary research, to always engage with an eye to the transformations of the contemporary city. «The aim of teaching urban archaeology is to gain an in-depth knowledge of the relationship between the archaeological traces of the cities that existed before the present settlement and the need to identify processes of urban transformation compatible with the layered and complex palimpsest of the historical centers» (Manacorda, 2008). Biddle and Hudson's *The Future of the London's Past*, (Biddle, Hudson, 1973) and Andreina Ricci's essay *Attorno alla nuda pietra* - "Around the naked stone" - (Ricci, 2006) are a first reference to approach the issue from the archaeological and conservative point of view. From our position of architects it is important that the study of the stratifications of the city's originary settlement systems and the acts of preservation that accompany any thinking on archaeology and the monuments do not ignore the form of the city. The architect's research will focus on respecting the rule of conservation in harmony with the form of the *urbis*, with the construction of the city and of the culture that it expresses. The city is a collective construction, based on its own history and representative of the people that built it; the city is always specific. Archaeology is the heart of thinking about the roots. Besides preserving the harmony with the form of the *urbis* and the specificities of the city, lies the issue of how to use archaeology's knowledge to plan the contemporary city without relinquishing the aesthetic aspiration.

Ever more often a consequence of an exasperated conservationist fundamentalism, or of inadequate cultural tools, has turned archaeological sites into archaeological enclosures, similar to the non-places, the islands of estrangement, described by Marc Augé (Augé, 2008). Architecture should, instead, address the ruin as a sedimented deposit of knowledge upon which to plan the new, regarding it for what it is in material and construction terms: the metamorphosis of an architecture, a part of all the historical architectures that comprise the architect's main tool of planning.

The enclosure and excavations connote the archaeological enclaves, constituting the main obstacle to grasping the connections that bind the ruins to the city, the soil's stratigraphies and the elevated constructions. Historical topography performs the necessary task of retracing possible sequences of representative spaces.

It carries on a patient research of the papers, the views, the representations that would help to grasp the connections and potential lines of continuities between the traces of the past and the city in becoming. The architect should prefigure a system of likely traces, connecting, integrating, entwining the archaeological ruins with the spaces of the contemporary city, where the red thread of history holds together multiple and multifarious elements, where the archaeological *insule* make up the foundation needed by architecture to pursue the value of the "eternal present" (Venezia, 2013). If this is the ultimate goal, we should create a methodology and establish a distinction between the archaeologist's "how" and the architect's "how", always keeping in mind that interdisciplinarity is one of the defining frameworks of such a project. Andreina Ricci argues that the first aim of the archaeologist is to help everybody to fully understand the meaning and value of the ruin and to virtually appropriate it by employing the two categories of "translation" and the "tale" that define "archaeological communication".

The architect's task is to implement the re-use of that asset in harmony with the "translation" and the "tale" envisioned by the archaeologist, never losing sight of the relationship between the form of the city and the architectural form of the ruin. Knowledge in its multiple meanings is the tool of the architect. As always happens with planning projects, critical orientation makes the work of knowledge essential and complex. Let us focus on three exemplary projects where the relationship of archaeology with the city and architecture is based on different modalities of articulating knowledge: Dimitri Pikionis' landscaping of the archaeological site around the Acropolis and the Filopappou Hill, Athens 1954-57, Giorgio Grassi and Manuel Portaceli Roig's restoration and rehabilitation of the Roman theatre of Sagunto, Sagunto 1985-93, and David Chipperfield's rebuilding of the Neues Museum, Berlin 1997-2009.

For Pikionis, the urban landscape enhances the knowledge of Athens' acropolis. The plan of the paths and platforms represents a realization of the dispositifs of vision of the acropolis. Pikionis practices a perceptive and material knowledge of the ruin in the landscape. «The Attic landscape, that has long lost its integrity, finds new life in the micro compositions that dot the path, in the narrow enclaves that give new meaning to the sense of the place, letting the original meanings ooze out of the rocks and intermingle with the other, century-old ones in the analogic play that regulates the succession of forms and ideas throughout history» (Ferlenga, 2006).

Giorgio Grassi rather adopts a mode of knowledge of the typological-constructive variant. This is knowledge of the constructive features of the Roman theatre, actualized by means of an accurate redrawing deployed by means of maps, sections, and prospects, of different Roman theatres, which he regarded as a necessary practice to reconfigure the absolute essence of the theatre's architecture. «The Roman theatre of Arles represents, from a material and especially symbolic point of view, the beginning of Grassi's Spanish projects between architecture and archaeology [...] The first concern that emerges from these notes is the notion that archaeology is integral to architecture and, as a direct consequence, the idea of 'ruin' as an architecture that lost part, or all of its original condition, without, for the same reason, relinquishing its typological and formal structure. It never ceased being an architectural form. Thus, the ruin is an architecture that has 'regressed' to a state preceding its completion, or, if we look at it from the opposite perspective, it is an 'unfinished' architecture that was interrupted halfway through its construction» (Malcovati, 2013).

Chipperfield practices a mode of knowledge based in the knowledge of the spatiality of "that ruin": this is a work on the architectural object in itself that reconfigures its perceptive unity. Chipperfield writes: «The process can be described as a multidisciplinary interaction between repairing, conserving, restoring and recreating all of its components. The original sequence of rooms was restored with newly built sections that create continuity with the existing structure. The almost archaeological restoration followed the guidelines of the Charter of Venice, respecting the historical structure in its different states of preservation» (Chipperfield's office web site, 2011).

In the enhancement project of the Archaeological Park of Baiae described in the following pages we mobilized the three modes of knowledge: the knowledge of the relationship that the ruin entertains with the landscape, the philological-typological-constructive knowledge, and the knowledge of the ruin's own spatial elements.

METHODOLOGICAL ISSUES

The method should be based on the principles of a theory that pertains to architecture, the project's sphere of activity, and aim at elaborating a process by establishing a common set of operating procedures for the projects within the different disciplines involved (architecture, archaeology, economics, as well as the institutions that deal with the protection and the local territorial governments). So an integral part of this process included the operations connected with dealing with the institutions that govern the territory to which the archaeological asset belongs; the governmental bodies in charge of its stewardship, the community of interested entrepreneurs, as well as with the state organizations which could offer funds to implement the development.

The process moves from a thorough understanding of the essence of what pre-exists. First of all an act of cognition is necessary. The same methodological consideration that invests the terms *archaeology* and *city*, but above all the interconnections established by these two terms, finds its inescapable beginning in the creative quest for knowledge. There can be no judgement without knowledge and there can be no project without judgement. The disciplinary tool which the architect mobilizes to know the city and architecture, or as one urban theory asserts, *the city itself as an architecture* (Rossi, 1966) is *Urban Analysis*, a tool which identifies specific epistemological modalities which depend on the urban project. In this disciplinary context, the project becomes an instrument of knowledge. If urban analysis provides us with a well-known procedure, in a certain sense codified by a reading of urban phenomena, for what, instead, pertains to the knowledge of the archaeological asset, we must necessarily refer to its material and formal consistency. It is clear that there exist several possible levels of knowing the archaeological asset. In light of this premise, the work involves a reading of the territorial scale, urban analysis, a formal reading of the archaeological asset, a research of historical documents, and a description based on a relief and re-drawing. After establishing the perimeter of the study-area, the urban analysis proceeds to include: an identification of the infrastructures system and accessibility; knowledge and description of the geomorphology of the ground and the configuration of the green areas and open spaces located in the analyzed context; a reading of the urban context (primary elements and construction pattern, urban morphology and construction typology); the investigation and reconstruction of the urban scenes on which the various configurations assumed over time by the archaeological site have referred to (documentary investigations through elaborate graphic descriptions). In contrast, followings are the tools that allow an approach to the knowledge of the architectural form of the archaeological asset: historical documentary studies; studies on architectural consistency, particularly material form and composite based principles; drawings from architectural surveys and reconstructions of the configurations assumed over time by the asset. The process's first step is identifying the founding elements of the asset's architecture, of the underlying structure of its formal configuration, to re-propose the recognized characters in the concept of the pilot project. In th s way a relationship of continuity and affiliation is established not with the linguistic and figurative apparatus of the asset, but with his *hidden structure* (De Fusco, 2000). An integral part of this process is represented by all the operations connected to the relationship with the institutions governing the territory to which the archaeological asset belongs; the governmental bodies with the community of interested entrepreneurs, with the state organizations which could offer funds to implement the development.

The complexity of this topic requires the cooperation of multiple actors to enhance the definition of the project, to transform the archaeological site in a propeller engine of the urban transformations of the local territory.

ABOUT THE RUIN

The enhancement project of the Archaeological Park of Baiae was developed under the scientific coordination of *Fondazione Internazionale per gli Studi Superiori di Architettura*, by the following team of architects: Uberto Siola, Renato Capozzi, Adelina Picone, Federica Visconti as senior scientists.

The monumental complex lays on the low part of the hill of *Baiae*. It consists of natural terraces, which, according to need, are moulded by powerful substructures acting as basis *villae*. It is a semi-annular shaped area of approximately 400 metres facing south-west to north-west. It is located between the Ferretti building and Piazza De Gasperi and features two nuclei, isolated from the original context by contemporary urban cuts (as shown in the so-called *Temple of Venus*), at the entrance of the harbour and the so-called Temple of Diana at the bottom of the modern square. The site is now under the authority of the Superintendence for the archaeological heritage of Naples and Pompeii. The first excavation in Baiae took place in 1800. In 1935 in obedience to Amedeo Maiuri's will, the Archaeological Park of the Baths of *Baiae* was created, following a process of land expropriation, excavations, restoration and safeguarding of a wide area expanding from the hill to the sea, which was however, enforced only in 1941. Other excavation and relief campaigns have been carried on in the area up to the present day, with the aim of restoring a more complete picture of the local topography. Experts have rarely agreed on dividing the site into sections, but today the complex is commonly to acknowledged as part of the imperial *Palatium* divided into the following sections: *Diana section, Mercury section, Villa Ambulatio, Sosandra section, Small Baths, Medium Level Baths, Hadrian Baths*. These designations are subordinated to description needs, as they refer, more or less, to autonomous architectural entities. The difficulty in detecting the use of these buildings is due to the network of structures and the presence of superimposed building phases that often prevent identification of the original architectural project of the buildings. The state of preservation of the area is not uniform and only a few original decorative coatings have been preserved (Guardascione, 2011).

Figure 1. The Archaeological Park of Baiae (Source: Authors).

THE AIM OF THE PROJECT

At the core of general theoreticals consideration and of planning experiences carried on by the project, the archaeological site should be the object of proposals concerning not only its restoration and preservation, but also its use in the present days and, therefore, its inclusion in the behaviour of the contemporary city in light of a wider territorial reorganization. All this invites to look at the archaeological heritage from the alternative perspective of architecture.

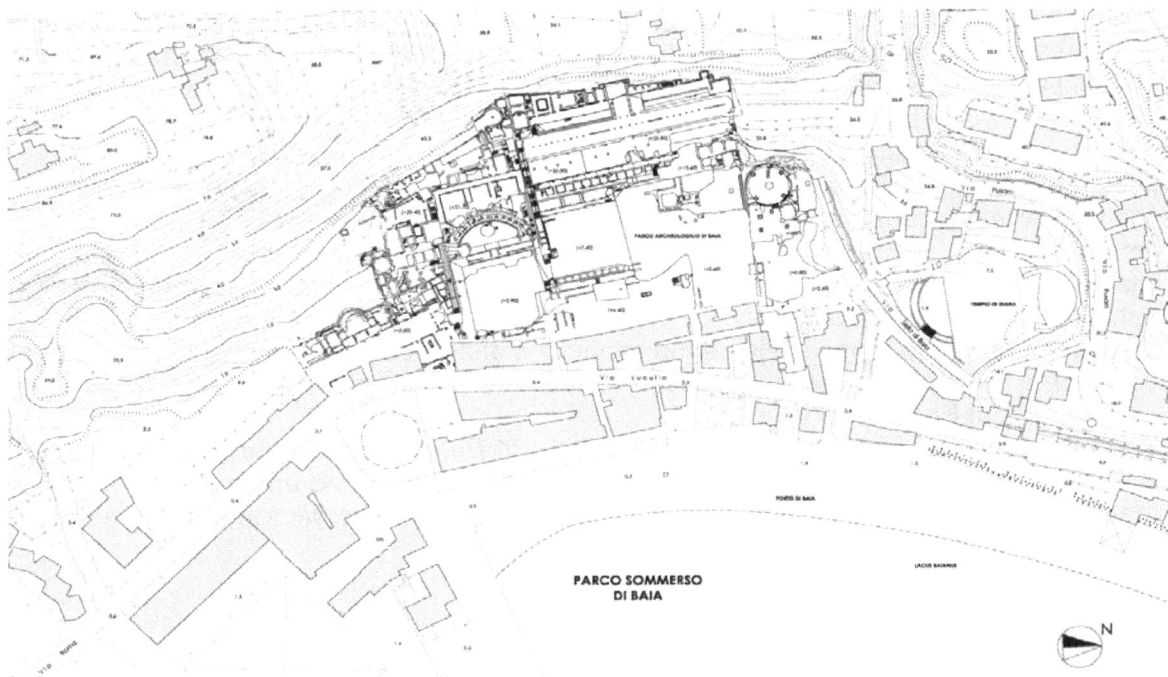

Figure 2. The Archaeological Park of Baiae: Consider The Monument,
redrew general plan of the ruin (Source: Authors).

Carrying out the project has meant broadening our vision to larger areas, other than merely the archaeological ones, which as a result of the identification of their own dimension, changed from study-areas into project-areas. Beyond attaching a significant value to the pre-existing archaeological sites, we worked within these areas to restore the functional, formal and meaningful relationships between the different elements involved. From the functional point of view, this meant working on accessibility and on the supply of services, while formally we also introduced new architectures, as we continued to settle and stratify traces. Finally as concerns meaning, we suggested to introduce new "themes" and identified new uses for the archaeological areas. The Archaeological Park of Baiae acquired the look of an area where the narration of architectural and landscape values merges with the theatrical one, in different forms, such as opera, classical and experimental performances, in harmony with the several spaces surrounding the complex. Furthermore, the project aims at reestablishing the original functionality of the ancient thermal baths, located in the so-called Mercurio Sector, where the architectural spaces are still well configured.

THE ENHANCEMENT OF THE ARCHITECTURAL DESIGN
The distinctive trait and, in our opinion, also the real value of our work consists in developing an approach to the archaeological issue that avoids creating physical and functional areas and fences. Thanks to its great value, archeology should instead be perceived as an area to protect, found within a broader "space" where changing opportunities can be seized by trying to turn actual remains, which are sometimes indecipherable except for the experts, into elements whose task is «to spread our present into the future» (Ricci, 2006). The only possible way to enact this spreading in an enhancement project is to achieve a great and deep knowledge. The analytical actions needed to acquire the knowledge of the archaeological asset have to be undertaken firstly: the archaeological research and the redrawing with reliefs to verify the correct representation of all the asset's portions.

The redrawing employed some reliefs obtained from the Superintendence as its basic plan, completed with measurements executed directly on site. The collaboration with the archaeologist was particularly relevant to obtain reliable sections and identify the right depth of the original levels. Redrawing can be considered a useful operation to attain a deep knowledge of the asset, introductory to conservation's actions. Further it acted as a sort of preparatory phase for the architectural design, particularly in the Mercurio Section, where great attention was paid to achieve the correct reconstruction of the original distribution and articulation of the spaces of the ancient thermal bath.

Starting from these premises, the design focused on two macro-themes, from which originated the following project topics:

- Improvement, to the urban scale, of the conditions of accessibility to the Archaeological Park, considered as a system within the cultural assets of the Phlegraean Fields in their totality, extending from Pozzuoli to Cuma.
- The new definition of the entrances to the Park of *Baiae*, identified according to the urban strategy Research of new functions with the goal of activating the enhancement's booster in the Archaeological *Park of Baiae*, considered as a part of the wide cultural system of the Phlegraean Fields.

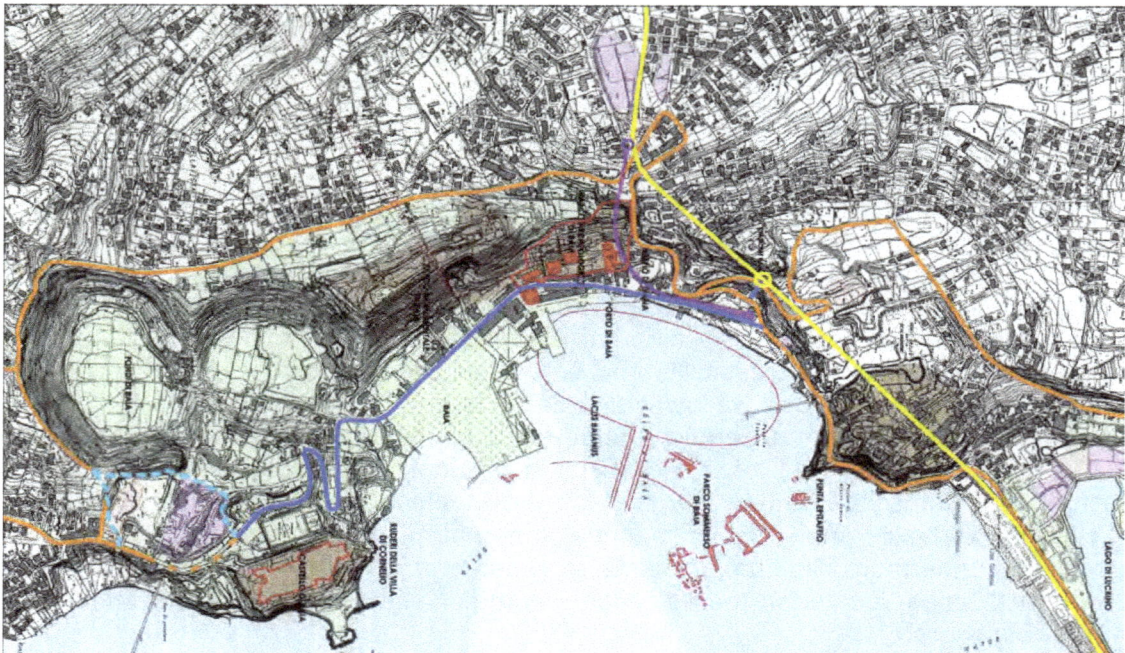

Figure 3. Accesibility and infrastructural system (Source: Authors).

Based on an architectural analysis, confirmed and strengthened by an economic-financial-management study conducted by UniMED, the possibility emerged that the Archaeological Park of *Baiae* could benefit from a new functional life. This stemmed, first, by considering its degree of accessibility, understood as an element of our cultural heritage, and secondly, taking a valuable action in full respect of the original configuration of the archaeological assets, implemented by restoring part of the thermal baths functionality in the Mercurio Section while using the open spaces as permanently-based venues of theatrical performances.

The project begins with an extensive study of the level of accessibility and infrastructural system of the archaeological monumental emergencies of *Baiae* (the archaeological monumental park, the submerged park, the *Castel of Baiae*) considering them as parts of a much wider complex of archaeological assets, which stretches up from Puteoli to Cumae and includes the Phlegraean Fields. The topics are:

- redefining the entrances;
- pedestrianizing the coast which supports the plan of accessing the Archaeological Park from the Temple of Venus, restoring the original relationship between the building and the *Palatium*;
- the new entrance from the Venus complex with a bookshop also accessible from an alternative route to the Archaeological Park;
- the entrance from the square directs visitors to the Mercury complex and implementing its restored thermal function;
- the established use of the open spaces as theatres, according to the various envisaged intervention levels: from the simple preparation of the seats and lighting fixtures in the so-called Temple of Venus for performances of experimental theatre, to the mounting of a "theatrical machine" in the *natatio* of the Sosandra section for performances of classical theatre and opera, as well as the layout of a natural *cavea* in one of the pensile garden for musical shows;
- the re-functioning of the baths in the so-called Temple of Mercury, foreseeing the restoration and a new functional life, closest to the original one, with the possibility to exploit the still-active thermal springs.

The development of these last two topics, related to the new functional life of the Archaeological Park, allowed to experiment with an interesting interdisciplinary approach.

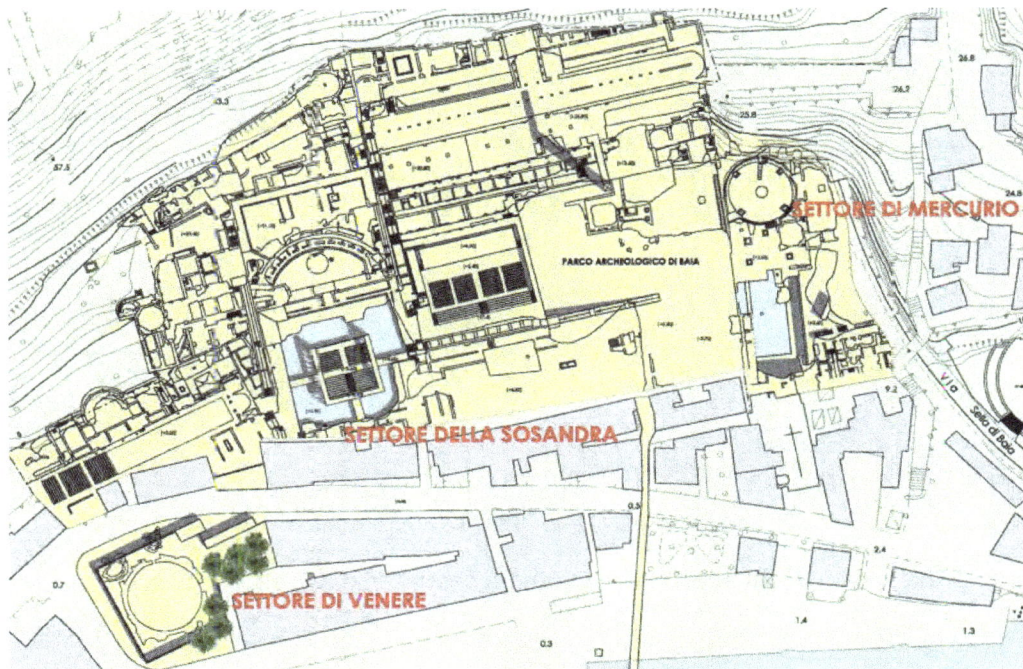

Figure 4. Archaeological Park of *Baiae*, the general enhancement project (Source: Authors).

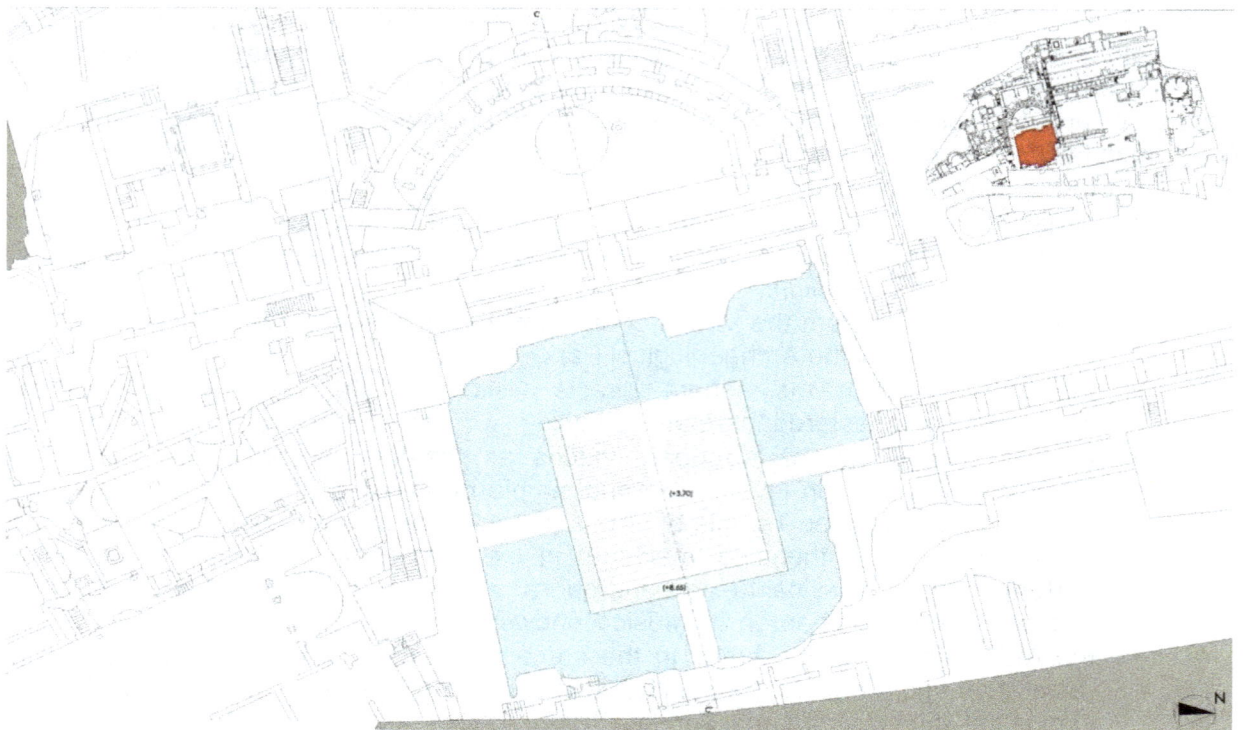

Figure 5. The project: the theatre in the Sosandra sector, plan and section (Source: Authors).

Figure 6. The project: the theatre in the Sosandra sector, plan and section (Source: Authors).

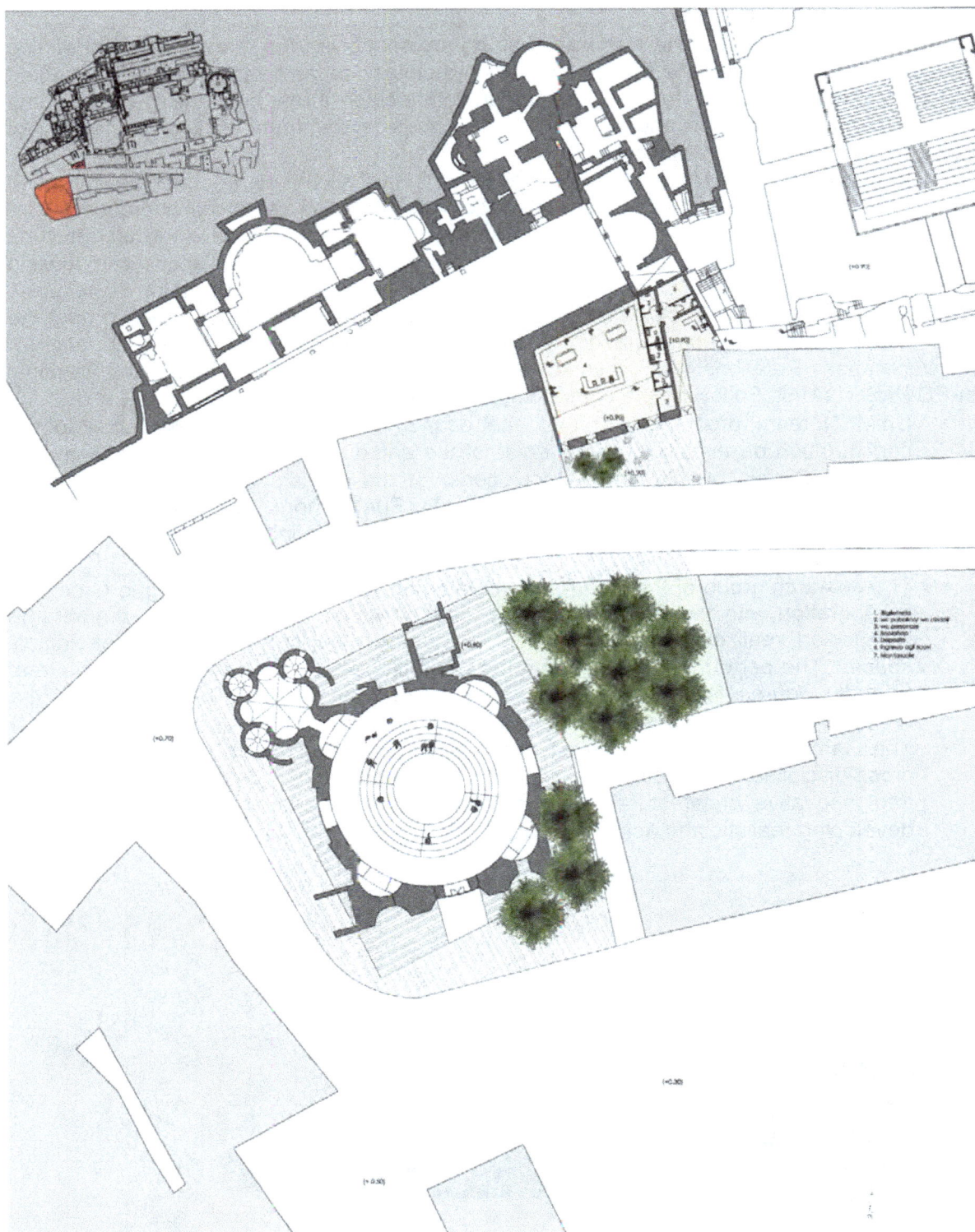

Figure 7. The project: the theatre in the Venus sector, plan (Source: Authors).

INTERDISCIPLINARY PROCESS

The different experts involved in the elaboration of the project played the relevant role of defining the work group, which required the coordination of an architect, an archaeologist - which worked not only as a consultant *ex ante*, but also as an interlocutor throughout the project -, and economists for context analysis and operational proposals in the framework of a strategy of revitalization and integration of resources.

It is really interesting to analyze this project in depth. In fact, all the elements of the explained methodology were epitomized in it. In the elaboration phase a number of collaborations were established with Flavia Milena Guardascione, an archaeologist involved also in the architectural project team; the team of economists composed by Luigi Manfra, Valerio Tuccini and Alessio Liquori, who worked under the scientific coordination of UniMED-Mediterranean Universities Union; a research group of the *Centro Interdipartimentale di studi per la Magna Grecia*, under the scientific coordination of Giovanna Greco; Campania Region's team, under the coordination of General Department Relations with National and International Bodies in matters of Regional interest-EU Projects Unit. Following the contributions of different actors:

- UniMED team produced not only the economic feasibility studies, but also the configuration of revitalization strategies of the entire urban environment, based on a deeper analysis of the context conditions, of the tourism market and of cultural asset's accessibility to the Phlegraean Fields. Furthermore, the economic research offered an important contribution to the new functions definition, as will be illustrated in the next sections.

- The research group of the *Centro Interdipartimentale di studi per la Magna Grecia*, in collaboration with the theatre company *Teatrocontinuo* directed by the dramaturge Nin Scolari, realized a theatrical experiment that was performed in the archaeological spaces. The performance, born from the scientific results of a research on the *Great Female Figures* from the ancient world, helped to test the potentialities of the archaeological park's open spaces.

- The Campania Region's team encouraged putting into practice the integration and cross-disciplinary contamination procedure, by envisaging comprehensive, cheaper and innovative architectural solutions and forms of management, with the aim of developing realistic and achievable action plans.

Figure 8. The project: the Mercurio sector Roman Thermal Baths, section (Source: Authors).

THE FIRST ENHANCEMENT RESULT.
The Archaeological Park as a multiform theatrical scene

The decision of holding a theatrical performance inside the Archaeological Park showed its potential function of events centre, considering the open spaces of the asset as background to all kinds of possible representations and performances: from experimental theatre to musical events, to grand opera theatre.

All the project's actors agreed with this vision, particularly the team of architects, which conceived the design with the aim to restore the ancient unity of the *Palatium*, highlighting the necessity of giving renewed centrality and urban visibility to the Archaeological Park. The new entrance, close to the Venus Temple, reconnects, also in a functional fashion, the monument with the Archaeological Park.

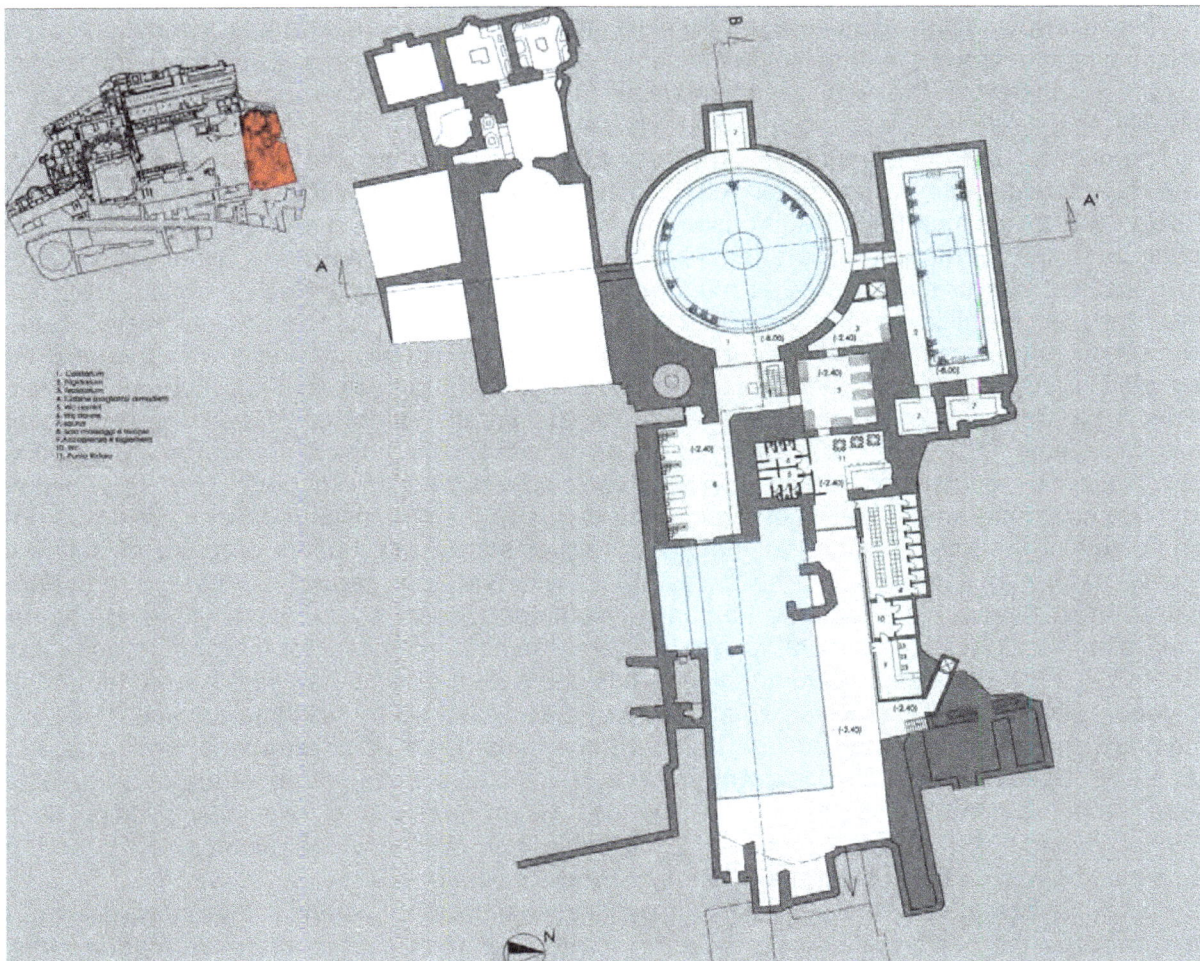

Figure 9. The project: the Mercurio sector Roman Thermal Baths, plan (Source: Authors).

The role of the Venus Temple is to reconnect and link once again the threads of the ancient tissue and the original landscape relations, providing a perfect chance to promote new urban centralitie in present-day *Baiae*. This is the meaning at the basis of the project, which intends to set the temple free from its fences and to restore its surrounding backspaces, re-adapting them for urban fruition. The expected interventions include the simple rationalization of the pavement and the green parterre of the garden that surrounds the monument, which, in the new functional life that ArcheoURB foresees for the Archaeological Park, will host experimental theatre performances and cultural events. The problems linked with the control, security and safeguarding of the archaeological ruins will be addressed by placing gates and glasses to the access rooms leading to the main hall, which will constantly be visible but accessible exclusively on planned occasions. Just in front of the Venus Temple, the project envisages a new access to the Archaeological Park using an area roughly based on via Lucullo, one of the few fissures of the curtain wall through which it is possible to actually see the archaeological complex. Here, we find the remains of the walls holding the above terrace of the Small Baths complex. These walls originally closed the halls, that according to archaeologists are of old age and were presumably used as warehouses. Here the designers plan to locate the bookshop and the ticket office, planning the front space in continuity and as an analogy with the parterre project of the Venus Temple.

This transformation was made possible thanks to the important contribution of the archaeological research, in order to give a certain knowledge of the original underground levels of the ruin. Furthermore the economical-financial study and the context analysis, conducted by UniMED team, were crucial in indicating the visibility-strategy.

The team of economists suggests to commit the crucial managerial aspect to a third subject, i.e. a foundation, who would be in charge of coordinating the different levels of expertise associated with the number of authorities involved: State, Region, Province, Phlegraean Park, Archaeological Government Department responsible for the environment and historical buildings.

In spite of the large funding that has been invested in the last ten years for the restoration works, for the rehabilitation and the environmental requalification actions, the expected cultural and touristic development in the Phlegraean Fields has never really started due to the lack of an integrated managerial action, and of the implementation of the system. The economic research makes a proposal for an operational solution for the enhancement of Baiae based on holding theatrical events within the Park: «the promotion of the Archaeological Park area by promoting cultural activities and events which may involve residents and day-trippers. The *Baiae* Roman baths is a charming area, open to different options due to the dimensional characteristics and the morphology of the site. The operational proposal devised in this project consists of setting-up several performance spaces in the baths area, with variable capacity, where a number of featured events (about 20) could be held in the summer season. Extraordinary funding channels could be employed (EU or national) for the facilities (the estimated cost for the performance areas is about € 4.2 m). Besides, regarding the events schedule (the estimated cost is about € 210,000 per year), grants for operating expenses, only partially covered by ticketing incomes, would be necessary and could be obtained from different channels (local administrations, sponsors, etc.). These activities could attract a relevant volume of visitors (a potential audience of about 8,200 people) and the contribution (the public and the private one) would be widely justified by the potential appeal of the activity, the cultural promotion of the archaeological area and the economic impacts on the retail trade» (Manfra, Tuccini, Liquori, 2011).

Following this idea all the *Palatium's* open spaces become potential theatrical spaces, as exemplified by the performance of the *Centro Interdipartimentale di studi per la Magna Grecia*, which chose an open space in the *Sosandra section* to host its performance, without requiring any sovra-structural installations.

«An extremely minimalist scenery has been requested and implemented, being almost inexistent in some particular areas where, indeed, it has been entrusted to the effects of lights, costumes and dancing, highlighting its emotional and fascinating side. On the whole, the scenery has never been invasive whereas the lights have emphasized the structures always present in the background. The monument, in all its parts, goes on playing the main character on stage and accompanies the audience towards the discovery of a beautiful niche, vault, or sinuous recess, which is its actual enhancement. The theatre stage, the scenery with its lights and sounds has never hidden the monument and the open-air Museum has gained great emotional momentum.

Figure 10. The project: the Mercurio sector Roman Thermal Baths, view of the new entrance

The scenery has made the best use of lights by re-using the shades of ancient structures, thereby giving rise to evocative powers and feelings that have never upset or effaced the ancient character and the original nature of the monument. At the end of this research activity, a story gathering texts, images, evidences, impressions discussed during numerous meetings and exchanges of views among the participants in the work group, philologists, archaeologists, historians of religions, comedians, has been staged. The story is set in the Mediterranean, sailed across at different times and along different routes. The performed travel is the travel of men, ideas, religions, cults and rites, which unfolds from the fall of Troy and the arrival of the Greeks on ancient Italy's coasts up to the present day. Great female figures that ancient tradition has passed on to us who answer the same strong and imperative appeal of motherhood. Guided by the strong royal power of Hera, protectress of sailing and happy landings, these Mediterranean goddesses-mothers narrate events and passions, changing routes, spaces and times, beyond rational, natural and temporal borders, thus becoming eternally universal icons and archetypes» (Greco, 2011).

Different and more complex is the case of a grand opera event that requires a large number of sovra-structural installations, as happened with the grand opera events (Cavalleria Rusticana, Aida) that the San Carlo Theatre of Naples staged in the Archaeological Park.

In order to stage effective theatrical activities in the archaeological park in terms of cultural promotion and economic impact, the offer of cultural activities should be diversified. Hence the need to host grand opera events without necessarily investing funds for the sovra-structural installations on every occasion, which would make the operation too expensive. The design meets these needs with its "theatrical machine". The new theatre - conceived as a demountable wooden structure - leads into the big court bringing it back to its original level and restoring its original function of *natatio*. The new building - as in the Roman tradition of "celebration machines" for *naumachias* or as for the maritime theatre of *Hadrian's Villa* - is an independent element inscribed as a fragment in the global additional composition for each single part, a new fragment which endeavours to making it easier for visitors to understand the rules and forms of the old times. In order to define its presence without imposing itself on the monument or competing with it, the theatre - from the architectural and syntactic point of view - thanks to an adequate and fit proportion, reaches the maximum height of the portico of the court; maintains in its overall layout the compositive axe passing by the centre of the exedra of the first terrace; and finally, due to its planimetrical form, gains the shape of a square. In this way, the theatre, isolated in its formal and dimensional individuality, stands out and reflects itself in the stretch of water linked to the portico by two pedestrian paths running on a piled structure facing east-westward which marks the passage between the two *cavea*, while a third path facing north-south links the theatre to the perimetric *deambulatio*, to the changing rooms, the warehouses and the lavatories on the ground floor. From a constructive point of view, the theatre is organized by a very high procession of slender pillars (in lamellate wood) constantly steady step, which, through the *vomitoria*, hold the slabs of the steps. These are completely pierced to increase the transparency of the entire building and, as a consequence, to consent the perception of the terraces of the *palatium* also in the backside towards south. This semantic choice makes the new building uncovered and open, attaching to the sole framing element - which joins together the two scene towers closing the orchestra - the definition of the architectural decoration whose main objective is to revive the general completeness, alluding in this way to a virtual cubic volume. The proper double inclination of the *ima cavea* and of the *summa cavea* guarantees a perfect visibility either of the stage of the maritime theatre, or of the complex architectural/backdrop system formed by a terraced system with concave and convex exedras of the ancient monument. The new building above whose composition, as mentioned above, is programmatically and evidently based on the old monument above proposes, therefore, a possible contemporary interpretation of the theatre theme, in the archaeological sites, basically as an "architectural device". Such an intervention would justify a new presence and collocation in a so delicate and complex context, acting as a suitable "watching machine" for shows and for the archaeological ruins without winking but with great respect and, at the same time, being able to glorify and acknowledge its beauty and value of testimony.

THE SECOND ENHANCEMENT RESULT.
Modern life for the ancient thermal baths in the Mercurio temple

As for the Mercurio section, the archaeological research ascertained the ancient function of the thermal baths, even if the original internal distribution of the spaces is not entirely clear due to the superficiality of the archaeological excavations. Most of the internal spaces of the Mercurio section are substantially still underground and their planking levels still unknown.

«As for the oldest complex we can only say that the big *rotunda* could be a *natatio*; here, drillings were done which reached a floor 8.50 metres under and intercepted a vein of water at 60° C. Chemical tests on water samples coming from the western area of the *rotunda* of Mercury were run, which have shown the presence of hyper thermal water with a temperature of 54,7° C, with organoleptic characteristics so to be classified as *strong sodium chloride* useful for balneo-therapy and mud therapy. The *Mercurio section*, which underwent magnifications, had to be very busy, but it never experienced functional alterations» (Guardascione, 2011).

The citation is from a Conference held in 1969 in the *Castello aragonese di Baia*, entitled: *Baiae Hydrothermal resources. Usages, Perspectives*, whose proceedings were published in 1997. The permanence of the water spring, the thermal proprieties of the water, the conservation of the thermal spaces, and mainly the striking *natatio* filled with water up to the vault, invite to imagine a new life for the *Palatium*'s Baths, not only for functional and financial reasons, but to restore the ancient *genius loci*.

It is almost unnecessary to stress the importance and effectiveness of reusing parts of the Roman baths of *Baiae* to promote tourist-economic development, not only for *Baiae* and the Phlegraean Fields, but for the whole regional territory. The main point is identifying the modalities of conceiving this reutilization which should be not only appropriate as regards the safeguarding of the archaeological heritage, but also an effective tool of preservation for two essential reasons. The first one is of strictly economic nature and concerns the possibility of entrusting the management of the thermal activity to a private company which will take charge of the ordinary maintenance of the complex. The second one, of educational and popular nature, concerns the appropriate usage of the original archaeological heritage which allows a more direct empirical knowledge, restoring the original function and finally making the Baiae Complex alive. The main theme of the project of the Mercury section is the excavation, dictated either by distribution or spatial choices. The excavation itself - about three meters under the actual level of the floor of the entrance of the so-called Mercurio Temple - which will portray the open space of rectangular shape, - compositive nucleus of the intervention: one space conceived as a water basin, a modern *natatio*, which will compose the archaeological findings that the excavation will bring to light - the natural forms of the landscape, the architectures (historical and contemporary shapes) - and which will also work as an access to the baths. The autonomous function of the baths will be ensured by a separate entrance. The entrance from *Baiae* square, designed and realized at the time of the last restoration works, was never used and has now been vandalized. The project foresees the same archaeological visit to the park, but starting from the new entrance located in front of the *Venus Temple* and it will be carried on by visiting the three sections, with the possibility to face inside the *Mercurio natatio*.

As far as the functional articulation of the spaces is concerned, around the central one formed by the big hall of the Mercurio Temple, the project intends to give back the original usage of the Roman baths to the halls and functions and foresees as well, in spite of a lack of reliable data which could probably be gathered following thorough and exhaustive archaeological researches, a destination compatible with the greatness and proportions of each hall and their proper use, in any case changeable, whenever new excavations bring reliable ascriptions to the punctual destinations, given the general sense of the operation.

The UniMED research highlighted the high touristic impact of this type of enhancement, with a really long range and huge potential. The main concern regards the managerial model, which would necessarily involve a qualified private actor, expert in the field. Following the formulated scenario: «It would represent an interesting opportunity even in terms of management because the maintenance of the area would be entrusted to qualified wellness and spa professionals, once spaces and structures have been restored and given their original function. In addition to maintenance and custody, this option may provide a potential income with positive relapses on the local economy (this initiative may take several legal forms that need further study and investigations). In the case under consideration, the particular features of the territory constitute a *unicum* which favours this union, by limiting to a minimum the harmful effects that are often attributed to the involvement of the private-profit sector in the cultural sector, and by providing an opportunity of "mutual support" thanks to which an important cultural heritage that is not available today may become accessible to the public. A time-regulated concession of the spaces can be hypothesized, through the payment of a fee in proportion with the income potential of the thermal activity. The resources deriving from this fee could be aimed at management of the part of the archaeological area that is open to the public (the possibility to start a project finance is not precluded). In order to understand the importance of the contribution that this solution could guarantee to the public mission of the supply of an available cultural heritage, some estimates and some basic, simplistic, yet crucial hypotheses in outlining the possible scenario, should be made. At this preliminary stage, according to some parameters gathered from field investigations, we hypothesize that the management of a spa facility in the Roman site may even produce a significant "compensation" flow aimed at the public management of the site, that may be approximately comprised between €100,000-€200,000. As stated, this hypothesis is very attractive and innovative but also actually applicable; it is particularly interesting because of the way in which it would make it possible to "open" a limited, yet important management space to private individuals, while still guaranteeing conservation of the heritage, sustainability of the cultural management and access to the public, in an area that today has a high risk of degradation and of being shut down» (Manfra, Tuccini, Liquori, 2011).

The interdisciplinary perspective from which we have elaborated this project requires the involvement of experts able to individuate the water springs still in action, to analyse the nature and properties of the water, and pushing away seawater seepages into the Mercury hall. Such a relevant research exceeds the competences of the project, which has anyway endeavored to bring to light, starting from a strictly urban view, the potential increased value in this field, giving a specific and clearly outlined scenario of an interdisciplinary methodology of research, waiting for funds that would allow to carry on with the plan.

REFERENCES

Augé M.(2008). Nonluoghi. Introduzione ad una antropologia della surmodernità, Elèuthera: Milano.
Biddle M., Hudson D. (1973). The Future of London's Past: a survey of the archaeological implications of planning and development in the nation's capital. Rescue publication n. 4: London.
De Fusco, R. (2000). La citazione in architettura. In Area, no. 51 July/August, 14-17.
Fazzio F. (2005), Gli spazi dell'archeologia. Temi per il progetto urbanistico. Officina: Roma.
Ferlenga A. (1999). Dimitris Pikionis 1887-1968. Electa: Milano.
Ferlenga A. (2006). Dimitris Pikionis ad Atene: cammini di pietra, recinti di sogni, Architettura di Pietra, Journal, http://www.architetturadipietra.it/wp/?p=194.
Francovich R., Manacorda D. (2000). Entry "Archeologia Urbana", in Dizionario di Archeologia. Laterza: Roma-Bari.

Greco, G. (2011). Madri/Terraneo. The Great female figures from antiquity. In Capozzi, R., Picone, A., Visconti, F. (Eds). ArcheoURB. Archeologia e città. (pp. 132-149). Clean: Napoli.

Guardascione, F.M. (2011). The Archaeological Park of Baiae. Short excursus on the archaeological research. In Capozz , R., Picone, A., Visconti, F. (Eds). ArcheoURB. Archeologia e città. (pp. 96-108). Clean: Napoli.

Irace F. (2011). David Chipperfield. Mondadori Electa: Verona.

Manfra, L., Tuccini, V., Liquori, A. (2011). Strategies for the integrated enhancement and management of cultural, environmental and tourist resources for the development of the Phlegraean Fields. In Capozzi, R., Picone, A., Visconti, F. (Eds). ArcheoURB. Archeologia e città. (pp. 114-129). Clean: Napoli.

Maiuri, A. (1930). Il restauro di una sala termale a Baia. In Bollettino d'Arte. Ministero della Pubblica Istruzione: Roma.

Maiuri, A. (1951). I Campi Flegrei. Istituto Poligrafico dello Stato: Roma.

Maiuri, A. (1983). Itinerario Flegreo. Bibliopolis: Napoli.

Malcovati, S., (2013). Architettura ed Archeologia: a proposito di alcuni progetti di Giorgio Grassi, Engramma rivista online, http://www.engramma.it/eOS2/index.php?id_articolo=1301

Manacorda, D. (2008). Lezioni di archeologia. Laterza: Roma-Bari.

Manacorda, D. (2009). Archeologia in città: funzione, comunicazione, progetto. In AA.VV. (Eds), arch.it.arch. Dialoghi d archeologia e architettura 2005-2006. Edizioni Quasar: Roma.

Matteini T., (2009). Paesaggi del tempo: documenti archeologici e rovine artificiali nel disegno di giardini e paesaggi. Alinea: Firenze.

Purini, F. (2000). Comporre l'architettura. Laterza: Bari.

Ricci, A. (2006). Attorno alla nuda pietra. Archeologia e città tra identità e progetto. Donzelli: Roma.

Rossi, A. (1966). L'architettura della città. Marsilio: Padova.

Settis, S. (2004). Futuro del "classico". Einaudi: Torino.

Segarra Lagunes M.M., (2002). Archeologia urbana e progetto di architettura. Gangemi: Roma.

Venezia F., (2013). Che cos'è l'architettura, Lezioni, conferenze, un intervento. Mondadori Electa: Milano.

AUTHORS

Renato Capozzi
Architect, Associate Professor
Department of Architecture_DiARC, University of Naples "Federico II"
via Toledo 402, Naples 80134, Italy
renato.capozzi@unina.it

Adelina Picone
Architect, Researcher and Assistant Professor
Department of Architecture_DiARC, University of Naples "Federico II"
via Toledo 402, Naples 80134, Italy
adelina.picone@unina.it

Federica Visconti
Architect, Associate Professor
Department of Architecture_DiARC, University of Naples "Federico II"
via Toledo 402, Naples 80134, Italy
federica.visconti@unina.it

5

A UNIFIED ARCHITECTURAL THEORY FOR ISLAMIC ARCHITECTURE

Sabir Nu'Man*
Ashghal Public Works Authority, Doha, Qatar

*snuman1978@hotmail.com

....

Abstract

This research aims to identify criteria for sustainable design solutions to meet the needs of Muslims today. Under the theoretical framework of Nikos Salingaros' Unified Architectural Theory, design needs and strategies derived from Islamic knowledge and values are identified and used to inform principles for building typologies, location, and movement between buildings. Green building technologies in line with Islamic values and examples of sustainability promoting policies are analyzed and used to further develop design strategies. The findings include a theoretical model that proposes the essential design criteria for appropriate architecture for the Muslim world. This research is relevant for architects designing for Muslims, who have a duty to create housing appropriate for their particular needs and cultural context.

Keywords: *Unified architectural theory; design theory; sustainable design; Islamic design; culture*

INTRODUCTION

Over 1.7 billion people, about 22% of the world population today, claim Islam as the guiding force within their cultural context. Of the total Muslim population, 97% are found in the developing regions of South and Southeast Asia, the Middle East and North Africa, and Sub-Saharan Africa (Pew Research, 2011). Throughout the Muslim world, a push towards modernization in the last 100 years has left little room for thoughtful consideration and application of climactic concerns or architectural heritage of the region. Architectural development was driven by the desire to either be in step with modern architecture on the side of Muslim architects or have modern architecture be cloaked in the decorative elements of Islamic heritage. Both of these extremes meant that the true spirit and heritage of Islamic architecture was neglected (Galdieri, 2002). The resultant surface approach to design reduces Islamic architecture to details that can be stuck on the surface of any building, undermining the opportunity for buildings to reflect design solutions that are representative of how Islam shapes the intended functionality of that building.

On a grander scale, the quest to understand and provide for the needs of Muslims today is the quest to find practicable solutions to political, sociological, and even intellectual dissonance that has occurred due to colonialism and other power struggles that have challenged the construction of cultural identity. As Mohamed Arkoun (2002, p. 7) writes, "we must recognize that the built environment in contemporary Muslim societies is under the influence of a generalized ideological bricolage which can also be described as semantic disorder". This semantic disorder or presence of competing building languages in combination with the disconnection between culture and the built environment requires a reinvigoration of the discussion about core principles of Islamic life that necessitate design solutions, historic technologies developed to address these, and new technologies that can be brought to bear on developing solutions. These considerations must form the core of an architectural curriculum in Muslim countries and be part of assessment criteria used by architectural oversight bodies (Benkari et al, 2011).

Hassan Uddin Khan (1982, p.15) writes, "when we finally find an intellectual basis for looking at this housing and defining our guidelines in those terms, then we will be able to create an architecture that is appropriate for Muslims, each in his own country". To address this question of finding an intellectual basis for defining housing guidelines, we can look to the *Unified Architectural Theory* posited by Salingaros (2014). According to Salingaros, modern architecture focuses on the form language, disassociating from the historical pattern, while traditional and adaptive architecture focuses on the pattern language and evolves the form from there. Here, pattern refers to how people interact with buildings through how they live their lives, and form is defined as "geometrical rules for putting forms together" (Salingaros, 2006, part 1). Salingaros offers a means of measuring whether a form language adapts to local context, and thus verifies his theory through its measurability. Developing *Unified Architectural Theory* further, this research works from the premise that both pattern and form are derived from Islam for Muslims. As the research shows, pattern comes from the particular needs for privacy and sustainability that comes from Muslim scripture and form comes from sacred geometry principles which help develop comfortable spaces.

Research Methodology

This study offers applied research aimed at producing sustainable design strategies for residential construction in the Muslim world today. Existing research has done much to identify sources of building codes in Islamic Scripture (Akbar, 1988; Hakim, 2006), applications of this code in history (Hakim, 2006; Ardalan et al, 1973; Marçais, 1928; Bianca, 1994), the environment and Islam (Abdul Haleem, 1999; Fazlul et al, 1999; Foltz et al, 2003; Izzi Dien, 1997), spirituality in architecture (Lawlor, 1982; Salingaros, 2014; Khan, 2015, 2013), sustainable architecture in Islam (El Wakil, 1981b; Fathy, 1986), and Islamic heritage and cultural continuity (Khan, 1982; Serageldin and Shluger, 2001), but these areas of research are rarely synthesized to develop real guidelines for architects today. Ultimately, this research aims to help architects and planners evaluate new residential developments and design appropriate solutions to house Muslims today. To this end, the research employs a qualitative and interpretive methodology, a combination of literature review, analysis of revealed doctrine, and examples of successful applications of sustainable practices. These will be used to identify design strategies that are derived from Islamic knowledge and values that shape different building typologies, their location, and the movement between them. Green building technologies are identified and discussed in context to the design strategies, while related examples of green policies will offer insight on how sustainable policies can be implemented in the Muslim world.

REVEALED DOCTRINE CONCERNING PATTERN IN THE BUILT ENVIRONMENT

The source of sustainable design as articulated for this research is revealed doctrine; what Allah and his Prophet (SAAS[1]) has said that helps to shape design decisions. Passages from the Qur'an and sayings from the Prophet (SAAS) that relate to design decisions, as well as the derived rulings based on those teachings are examined. To help structure the discourse, three major spaces in the urban fabric are defined: homes, streets and the placement of building types that shape movement within a community. These design principles have been developing for hundreds of years and still have currency today (Hakim, 2006).

---

[1] SAAS here means Salallau Alayhi wa Salaam or "Peace and blessing be upon Him", an honorary invocation Muslims give upon hearing the Prophet mentioned.

Home: The mini mosque

Homes can be looked at as the second most important typology in the community after the mosque. Allah has said about the mosque, "The masjids are for Allah, so do not invoke anyone along with Allah" (Al-Qur'an 72:18). In other words, its function is to be a safe place for people to worship Allah; and worship, of course, is paramount: "We did not create jinn and men except to worship Us" (Al-Qur'an 51:56). Homes share an analogous function to the mosque, but on a smaller scale. Homes are to be safe, clean, and protect its inhabitants from harm while doing no harm to others.

Home environment: Respect for site, safe and healthy materials

In order for a home to be "safe", as envisaged in the Islamic paradigm, it must respect the environment in its site placement and building materials. Allah orders mankind not to commit abuses on the earth by saying, "Eat and drink from the provision of Allah, and do not commit abuse on the earth, spreading corruption therein" (Al-Qur'an 2:60). He also says, "And do not seek corruption in the earth; for God likes not workers of corruption" (Ibid, 28:77). To add to this awareness of how important it is to not spread corruption, the Prophet Muhammad (SAAS) said, "He who cuts a lote-tree [without justification], Allah will send him to Hellfire" (Abu Dawud). He also said that animals are worthy of protection, saying "There is a reward for serving any living being" (Al Bukhari, 1997). From these teachings, it is clear that protecting the environment is a duty, and in doing so there is a reward. Architect Hassan Fathy (1986, p. 5) wrote regarding the responsibility of the architect:

> "He is introducing a new element into an environment that has existed in equilibrium for a very long time. He has responsibilities to what surrounds the site, and, if he shirks this responsibility and does violence to the environment by building without reference to it, he is committing a crime against architecture and civilization."

Home design: Privacy through windows, doors, and courtyards

From the teachings of Allah and his Prophet (SAAS), we have both direct and indirect guidance around construction. Concerning the protection of privacy, Allah says, "O you who believe, enter not into houses other than your own until you have asked permission and have properly greeted those in them" (Al-Qur'an 24:27). The ways a person can enter and observe what is happening in someone's house are by looking through the windows or doors of the house. Allah also commands believers, "Say to the believers that they should lower their gaze and guard their modesty" (Al-Qur'an 24:30). Inspired by these teachings, major design strategies for homes have been developed in the Muslim world, such as the interior courtyard and the bent entrance.

The interior courtyard allows sunlight and air circulation into the home without exposing the private life of the house to people passing on the street. When it is unavoidable to have a window that faces the street, mashrabiyas (or latticed screens) can go over the window. Mashrabiyas have different patterns formed by adjusting the size of the spaces between and diameter of the balusters have the following functions: "(1) controlling the passage of light, (2) controlling the air flow, (3) reducing the temperature of the air current, (4) increasing the humidity of the air current, and (5) ensuring privacy" (Fathy, 1986, p. 47). Also, raising the window above eye level is used as another solution. The bent entrance allows people to open their door and enter the house without exposing the inside of the house to people passing on the street.

These design strategies allow the home to comply with the teachings of Allah and his Prophet (SAAS). They also reinforce the Prophet's (SAAS) statement that "there should be neither harming nor reciprocating harm" (Al-Muwatta, 1982). In the context of home design, harm is when someone outside can observe private behavior inside the home. Those who do look without permission can be punished without recrimination. The Prophet (SAAS) said: "He who looks into the homes without the occupant's permission, and they puncture his eyes, they have no right to demand a fine or ask for punishment." The Prophet (SAAS) also said, "Do you know the rites of the neighbour? You must not build to exclude the breeze from him unless you have permission" (Ibn Adi, 1938). This saying from the Prophet (SAAS) along with the others that we mentioned inform policies that help shape the built environment within the Muslim community. For example, Saydi Umar (RA) ruled on a case in which an upper floor addition had a window that overlooked the neighboring property. Umar requested that someone step on a bed and look through the window; if he saw what was in the neighbor's house, the window should be sealed (Akbar, 1988, p. 225-6). Ibn Wahb (d. 813 AD) applied the same rule to doors (Ibid, p. 95). The location of the windows and doors is therefore a major design decision that can determine whether or not a home complies with Islamic principles of privacy.

Home conservation of resources: Don't waste water, energy, or materials

Beyond the external privacy concerns, there are also prescriptions for the internal environment that have an effect on the external environment and other building typologies. For the external environment, a crucial understanding is that the earth was given to man as a trust. Allah says, "Indeed, I will appoint upon the earth a vicegerent" (Al-Qur'an 2:30). The common understanding is that man should behave as a representative of the way of Allah on earth, implying that man is a reflection of what he was created to represent. The way of Allah is to act according to Allah's will and maintaining Allah's limits. For example, he says "Eat of their fruit when they are in season, and pay their due upon harvest day, and waste not, for God loves not those who waste" (Al-Qur'an 6:141). Here it is clear that wasting resources is displeasing to Allah, and the Prophet (SAAS) said, "Do not waste; do not use more water than you need" (Ibn Majah 1, p. 34). To help prevent the home from wasting energy, windows and doors should be put in areas that will help reduce the energy use of the home while still performing the appropriate functions. The sun can be used to help cool the home through evaporation and provide energy through the use of solar panels.

Street: Size and provision for waste management

Instruction around the minimum width and the behavior of the street also comes directly from the Prophet (SAAS). Concerning the width, he said "If you have a dispute about the limits of the road make it seven cubits and then build" (Al-Bukhari, 1997). Seven cubits is about three meters, approximately the minimum width of one lane according to the National Association of City Transportation Workers (2016). This statement helps to provide a starting place for street organization. Umar b. al-Khattab advised in relation to organizing the streets in towns being founded in modern day Iraq, that "main roads (be) thirty cubits, and those in between twenty; lanes [aziqqah] seven, and the alley [fiefs'] width or length sixty cubits" (Akbar, 1988, p. 85-6). This can be looked as the foundation of the movement within the Muslim city which creates a semi-private, semi-public, and public layout that can still be seen today in cities like Fes, Algiers, Tunis, and Damascus.

About behavior in the street, the Prophet (SAAS) emphasized cleanliness, saying, "Avoid three things accursed: excreting in streams, thoroughfares, and in the shade" (Abu Dawud, 2008). Since "thoroughfares" here refers to the roads, the hadith pertains to one's behavior when using the streets, which are public assets. He also said, "If a man is walking in the street and finds a branch of thorns and removes it, then God will thank him and forgive him" (Ibid). Cleanliness is a quality that is close to Allah which we can understand from the Allah's statement "Truly, Allah loves those who turn unto Him in repentance and loves those who purify" (Al-Qur'an 2:222) and the Prophet's (SAAS) statement, "Keep yourselves clean as Islam is clean" (Ibn Habban, 1949). He also said, "God (be praised) is good and loves goodness, clean and he loves cleanliness, generous and he loves generosity, perfect and he loves perfection, so clean your fina" (Al Tirmidhi). The term fina means the outside property between the home and the street.

According to Imam Malik (711-795), this area is owned by the abutting property owner. The second caliph Umar proclaimed that the *fina* belongs to the house owner whether it was on the front or the back of a property (Akbar, 1988, p. 109-10) (See Figure 1). To help the community comply with the instructions from the Prophet (SAAS), spaces and streets must be designed in ways that make trash collection and recycling, for example, easy and thus compliant to cleanliness.

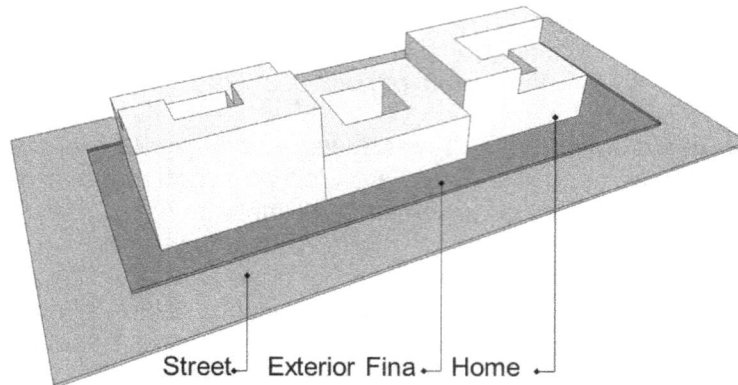

Street⌐   Exterior Fina⌐   Home ⌐

Figure 1. Exterior fina of the homes in relationship to street (Source: Author's Drawing).

Community: Easy access to mosques, markets, and schools
The third realm of the built environment involves other building typologies that make up and serve the community. The word that is used to represent community and the infrastructure that supports the community is *medina,* an Arabic word created from the root letters *mim*, *dal* and *nun,* whose derived words refer to civility and decorum. The word *medina* contains an implied concept of order that governs the infrastructure of the community. For a community to be considered to be a *medina* there are specific typologies that should exist within the community. Hakim (1988) quotes from Marçais' 1928 article "Islam and Urban Life" that "Malik, the father of the Maliki School of law, recognized a Mesjid al Jami - the mosque in which the Friday noon prayer and Khotba is undertaken - only in those settlements which had a *Suq*" (p. 57). This implies that there are three basic typologies. First, the Masjid where the Friday *khutba* is given implies that there must be someone with the appropriate knowledge to perform the *khutba* correctly as the Prophet (SAAS) did. Second, there must be a market (*Souq*). The third typology is the home, which also implies that there are streets to connect the different typologies to help people move through the spaces.

An additional aspect related to the shaping of the community is found in Surah al-Jumu'a when Allah says "O ye who believe, when the call is proclaimed to the prayer on Friday, hasten earnestly to the remembrance of Allah and leave off business, that is best for you if you but knew. And when the prayer is finished, then you may disperse through the land" (Al-Qur'an 62:9-10). It can be deduced that because of how the verse speaks of the relationship between the place of prayer and the place of business, it is important that these two spaces have a clear and easy flow between each other. This can be seen in most Islamic cities and they are looked at as the heart of the overall community (Hakim, 2006, p. 69; Bianca, 1994, p. 36).

PATTERN AND FORM FROM SUSTAINABLE STRATEGIES APPROPRIATE FOR MUSLIMS
As was found in the previous discussion on scripture related to design, conservation of resources is a central tenant of life for Muslims. This section addresses home solutions to prevent waste in the three topical areas of water, food, and energy by citing existing examples where this has been effective and correlating it with its related scripture.

Water: Prevent waste

Water is considered one of the most important components of life. Allah says, "Allah has created every [living] creature from water" (Al-Qur'an, 24:45). He also says, "He who made for you the earth a bed [spread out] and the sky a ceiling and sent down from the sky rain and brought forth thereby fruits as provision for you" (Al-Qur'an, 2:22). This verse speaks about both direct sustenance from water through drinking it and indirect sustenance through eating plants and animals. In another verse Allah says, "We have sent the fertilizing winds and sent down water from the sky and given you drink from it. And you are not its retainers" (Al-Qur'an 15:22). The portion of the verse, "And you are not its retainers," is a reminder that water is not personal property. The Prophet (SAAS) said, "On the day of resurrection God will not consider or support and will make a man face severe torment who had access to water in a thoroughfare and denied it to passers-by" (Abu Hurairah). All of these statements illustrate how important it is to protect this resource.

The Prophet (SAAS) emphasized how important it is not to waste water. When the Prophet (SAAS) saw Sa'd performing wudu and wasting water he said water can be wasted during wudu "even if you perform it in a flowing river" (Ibn Maja). Water can be misused even when there is an abundance of it, and even more so where it is scarce. Today, water recycling helps to protect homeowners from misusing water, because it takes the water that is used and cleans it for reuse. This can be done on a large scale such as with wastewater treatment plants (Ashghal, 2014), or limited to single home grey water recycling. Residential grey water recycling is safe and effective. Currently in the United States, for example, "there are eight million grey water systems with 22 million users. In 60 years, there have been one billion system user-years of exposure, yet there has not been one documented case of grey water transmitted illness" (Ludwig, 2006).

Waste management and composting: Reduce and reuse

How a city disposes of its waste s a major factor in how sustainable it can become. For all cities, dealing effectively with waste from new consumption patterns entails reviving rural practices of composting in an urban context, and instituting recycling collection systems. Caring for the earth is an edict from Allah who has said, "It is He who hath produced you from the earth and settled you therein" (Al-Qur'an, 11:61). Treatment of the earth can be seen as a reflection of how man treats himself. In order to correct the current environmental conditions, humans must change their own condition, for Allah has said, "Verily never will Allah change the condition of a people until they change what is in themselves" (Al-Qur'an 13:11). The spiritual cleansing implied by "change the condition" in this verse has both an internal and external reality, an internal rehabilitation, the external manifestation of which is an improved treatment and rehabilitation of our environment.

Composting food waste is integral to rehabilitating the environment, helping to replace the important minerals that food production takes out of the earth. This also helps to strengthen the earth for future generations, continuing the legacy of sustainability established by Prophet Muhammad (SAAS) who said, "Even if you fear that the Last Day has arrived, plant the sapling you hold in your hand" (Al-Bukhari, 1997). Therefore implementation of recycling trash and composting food waste can be seen as necessary to help reinforce Islamic values which advocate caring for the environment regardless of the situation. Cities like San Francisco and Oakland, both in California, USA have adopted mandatory recycling and composting ordinances that require residents to separate their recyclables, compostables, and landfill trash. Comprehensive citywide recycling was pushed through legislation by the organization Green Cities California (GCC) whose mission is "to accelerate local, regional, national and international adoption of sustainability policies and practices through collaborative effort" (Green Cities California, 2013). Sustainable development can be achieved by educating the public and providing infrastructure. Public infrastructure would include environmentally friendly design for homes and neighbourhoods and compost and recycling collection services.

Energy: Solar, thermal mass, material selection, and strategic design

In Islam, conservation of energy and use of clean energy is important. Allah says "And He hath made subject to you the sun and the moon," (Al-Quran, 14:33). Allah gives explicit permission to use the sun to meet human needs and enhance the quality of life. Solar panels are designed to absorb the sun's rays and convert them into electricity or heating. The panels capture the sun light through a silicone system and transform it into direct current electrical power (One Green Home, 2013). They can be incorporated into neighborhoods in a number of different ways. For homes, they can create more shade for the roof and provide renewable energy for the homes. Each home within a neighborhood and its corridor can be connected to a local energy grid where the excess energy produced could power the adjacent non-domestic buildings and provide energy to light the streets. The repetition and connection of these neighborhood power grids can be joined and the excess energy produced can be used to power buildings in public spaces, such as mosques, schools, and markets. By introducing solar panels, a new industry built around their production and maintenance would also be created, helping to both free communities from dependence on non-renewable energy sources and increase community sustainability.

Reduction of energy consumption in the built environment can also be achieved through selecting appropriate building materials based on the environment of the site and reducing the use of the cars in the neighborhood. The selection of appropriate building material is vital in maintaining a comfortable temperature for the users of the building. The ideal temperature is from 68°F (20°C) to 78°F (25°C). To create this temperature inside a sealed building made of concrete and steel, builders rely on heating, ventilation, and air conditioning (HVAC) for up to 9 months of a year in areas of extreme heat in the summer and average temperatures below 69% in the cold season, such as the Arabian Peninsula (Weather Underground, 2013).

Thermal mass can help reduce the amount of energy needed for the building to maintain a comfortable temperature with less reliance on these energy-reliant systems. Thermal mass is when the mass of the building provides "inertia" against temperature fluctuations outside of the building helping to maintain a comfortable environment inside. Hassan Fathy (1986) found that thermal mass using mud brick is a better solution than reinforced or prefabricated concrete to create a comfortable environment within a building. This was shown to be true through a 1964 experiment by the Cairo Building Research Center, where two small buildings were built, one from prefabricated concrete with 4 in (10 cm) walls and the other from mud brick with 20 in (50 cm) walls. The inside temperature for the prefabricated concrete building reached 97°F (36°C) when the temperature outside during the day only reached 82.4°F (28°C) adding about 15° degrees of heat higher than the temperature outside. Inside the mud brick building, temperatures reached only 73.4°F (23°C) which is within the human "comfort zone" (Ibid, p. 40). This shows that the mud brick building with 20-inch (50 cm) thick walls requires less energy to maintain a comfortable environment than standard concrete construction.

There are other materials that can be used to create thermal mass, such as straw bales which create a wall mass between 18-23 inches (46-58 cm) and rammed earth with walls between 12-24 inches (30.5-61cm). For countries with lime deposits and high heat, such as those on the Arabian peninsula, the most appropriate thermal massing building material is sand lime (calcium silicate) bricks which are made from lime sand and water. After formation, the bricks are then hardened in autoclaves under steam pressure at temperatures between 160°C and 203°C. Lime and quartz sand react to form compounds that give calcium silicate bricks their compressive strength (Making Sand Lime Bricks, 2012). This material can be used to create walls that are thick enough to create thermal massing. Because this material is natural to the environment, it will return to the earth in a clean way unlike concrete and steel rebar which are the predominant construction materials in the Muslim world. Using locally sourced material has the potential to reduce the Muslim world's dependence on concrete, providing more options for appropriate building materials while maintaining the pattern language of the region.

Another area in which the Muslim world can achieve greater energy efficiency is to reduce dependence on cars by designing all the major typologies that are needed for the community to function within a one-mile radius. This means that people can leave their homes and walk to the market, school, or green spaces within 20 to 30 minutes. A new city proposal that will be built outside the city of Chengdu, China does just this. Designed by Adrian Smith and Gordon Gill Architects (2012), it aims to accommodate 80,000 people, using 320 acres (1.3 km²). Laying out the city for people to be able to walk to get the things they need is not a new phenomenon within urban planning.

FORM VIA SACRED GEOMETRY

Shape and proportion affect our sense of order and wellbeing and are therefore an important element of design. In addition, shape and proportion can be measured and studied to help determine its appropriateness for different typologies (Salingaros, 2006). Certain proportions are repeated by Allah because they reflect the divine oneness and are a means of communication between the seen and unseen. Afdal al Din writes (in Nasr, 1968 p. 296), "The physical world is the symbol and image of the spiritual world". Al Ghazzali adds (in Smith, 1944, p. 111), "The visible world was made to correspond to the world invisible and there is nothing in this world but is a symbol of something in that other world". For mankind, these symbols are both natural and revealed, and have a transformative effect on hearts and minds. "Symbolic form, which are sensible aspects of the metaphysical reality of things, exist whether or not man is aware of them - man does not create symbols, he is transformed by them" (Ardalan et al, 1973, p. 5). Effectively, the symbolic is a means of communication that exists between the seen and unseen and has the power to shape our hearts and guide our thoughts and movements. Because of the power of this communication between these two dynamic realities, there is a need for guidance to help protect the one who shapes the physical world and to protect the physical world from the one who shapes it.

This guidance comes in the form of signs highlighted in the Qur'an, where Allah explains "Soon will we show them our signs in the (furthest) regions (of the earth), and in their own souls, until it becomes manifest to them that this is the Truth" (Al-Qur'an, 41:53). He directs us to look at the creation to see those signs by saying "He Who created the seven heavens one above another: No want of proportion wilt thou see in the Creation of (Allah) Most Gracious. So turn thy vision again: seest thou any flaw?" (Al-Qur'an, 67:3). These verses are touchstones from which to derive decisions and choices that do no harm to the earth or humanity. From these statements, Allah is directing humans to look at the earth to help increase understanding of the self. The Prophet (SAAS) said, "He who knows himself knows his Lord," (Al-Sakhawi) for knowing one's Lord is the key to achieving the reason for existence.

Through geometry, which means "measure of the earth" (Lawlor, 1982, p. 6), many connections are made between man and the natural world, one of which is the proportion with which physical forms are created known as the golden proportion.

The Golden Proportion is a constant ratio derived from a geometric relationship, which, like π and other constants of this type, is irrational in numerical terms. In a sense, the Golden Proportion can be considered as supra-rational or transcendent. It is actually the first issue of Oneness, the only possible creative duality within Unity (Ibid., p. 46).

This Golden Proportion also can be looked upon as the foundation on which the natural world is built and continues to develop. Through it, a harmony and physical rhythm continues to manifest, protecting it from any flaw, hence Allah's statement, "So turn thy vision again: seest thou any flaw" (Al-Qur'an 67:3)? From these sources are two realities that will help guide the design process; the laws of Geometry that are the foundation for physical beauty, and the divine Knowledge from Allah and his Prophet (SAAS). With these two realties, an environmental balance can be developed that will enhance the existing life patterns.

It is useful to examine some examples of architecture that have used the two realties mentioned to help shape their design dimensions. Hamdy house in Cairo, Egypt, built by Abdel Wahed El-Wakil in 1978 is one such example. El Wakil said that "because of the nature of the volume, considerable use of geometry and arithmetic proportions, such as the Golden Section, was made in order to give intelligible scale and proportion to the composition of spaces" (El-Wakil, 1981, p. 58). It is clear from the floor plan how he used the Golden Proportion to help organize the space. He also applied divine knowledge in relation to protecting the privacy of the homeowner from passersby. He did so by placing the windows above eye level and orientating the house around an interior courtyard enabling the inhabitants to move freely, enjoying the outside and inside spaces, all the while protected from onlookers passing by.

Another example can be found in Jami Mosque in Isfahan, Iran, finished in 1367, which is described as follows:

"The curvature of this interior is such that a pentagon is generated between the sides of the equilaterals and the produced arms of the re-entrant angles. Since it is the property of the pentagon that is perpendicular from the apex to the base is divided at its Golden Section by a line joining the remaining angles." (Pope, 1965, p. 1008, in Ardalan et al, 1973, p. 23)

The pentagon "reveals the relationship of $\sqrt{5}$ both with the number 5 and with the fivefold symmetry of the pentagon" (Lawlor, 1982, p. 36). Lawlor also mentions,

"The Golden Proportion generates a set of symbols which were used by the Platonic philosophers as a support for the ideal of divine or universal love. It is through the Golden Division that we can contemplate the fact that the Creator planted a regenerative seed which will lift the mortal realms of duality and confusion back towards the image of God." (Ibid., p. 46)

The Golden Proportion is also used in the tomb of Imam Ali Badar al-Qarafi (completed 1310) in Cairo, Egypt. With these examples we can see two realities: first, the divine Knowledge from Allah and His Prophet (SAAS) and second, the laws of Geometry that are the foundation for divine beauty that inspires the souls of witnesses.

The effect of the use of the golden proportion in these buildings is in stark contrast to attempts by Le Courbousier in his Unité d'Habitation and Neufert's *Bauentwurfslehre* which both claim to use the golden proportion and mean. As Frings (2002) points out, contrary to letting sacred geometry have a guiding influence on their work, "in fact, Neufert and Le Corbusier seem to use the Golden Section as a way to embellish their own subjective artistic creation by theory and ratio" (p. 31).

In practical application today, the golden ratio can be used in sizing building floor plans, facades, and master plans for residential developments. Its basis in human and natural scale helps ensure that the form fits into the overall human patterns of existence, and by expressing a link to the infinite in a finite form, the space will help its occupants connect to their creator, the very purpose of life for Muslims around the world.

ADAPTABILITY TO PLACE

In applying the design strategies pulled from Islamic scripture and developments in sustainable building practices that dictate the pattern or way people interact with space and sacred geometry as the source of geometrical rules for putting forms together to a specific place, it is necessary to add consideration of the region. According to Salingaros (2006), regional considerations are both part of the form language in that they "arise from available materials" (p. 1) and are a consideration when combining form and pattern in his adaptive design method. Here we differ, finding through sacred geometry that proportion is not affected by materials, although materials themselves do play a role in adaptive design. Design for place is the result of the combination of form, pattern, and regional sensibilities to meet the needs of people. For example, Figure 2 shows interior courtyards, a specific pattern cropping up in different regions with different climactic conditions; Northern Morocco on the left and Egypt on the right. The details reflect the different regional sensibilities of the two locations. For Morocco, the language relies on Zilig tiling, which covers the floor and extends up the wall. Wood lattices and doors and plaster work complete the look. For Egypt, the language of carved stone and stone lattices is used to beautify the spaces. Tiles are used for the floor only and wood lattices and doors are finishing touches. Decorative arches and columns are used in both examples, but they are expressed differently. For Salingaros (2014), this design language can be studied using the *Unified Architectural Theory* form language checklist and he posits that complexity correlates with its level of regional adaptation. From this, it is clear that careful study of regional sensibilities should be made and used to inform the language of the finished product.

Figure 2. (Left) Courtyard home in Fes, Morocco, (Right) Courtyard home in Cairo, Egypt (Source: Author).

CONCLUSION

This research provides architectural guidelines and policy suggestions rooted in the Qur'an and Sunnah that can have an immediate and lasting effect on the built environment in the Muslim world. The findings are in keeping with *Unified Architectural Theory*, which articulates that architecture must be in keeping with both form and pattern languages in order to create truly adaptive architecture. By showing how the embedded knowledge found in revealed doctrine and the teachings from the Prophet Muhammad (SAAS) is part of a Muslim's universal pattern language and how Allah's creation provides guidance for the form language, it posits that this guidance is relevant in leading communities toward a sustainable future. These design principles are critical in helping to revive Islamic knowledge with regards to its ability to help shape the built environment. Islamic architecture goes beyond surface details and penetrates to the heart, not only of the overall community layout, but also the specific typologies that come together to form the layout.

The design strategies extracted from the Qur'an and Sunnah are adaptive and can be applied to any region in the world to help produce Islamic architecture that is appropriate for that regional climate and its culture. The four elements (See Figure 3) used to help guide the design process for Islamic architecture are therefore:

A. The divine Knowledge from Allah and His beloved Prophet (SAAS) (i.e. pattern),
B. Sustainable strategies and technologies appropriate for the site (i.e. pattern and form).
C. The laws of Sacred Geometry (i.e. form), and
D. The regional sensibilities which include the environment, aesthetics, materials, and historical influences on the people in the region (i.e. adaptability).

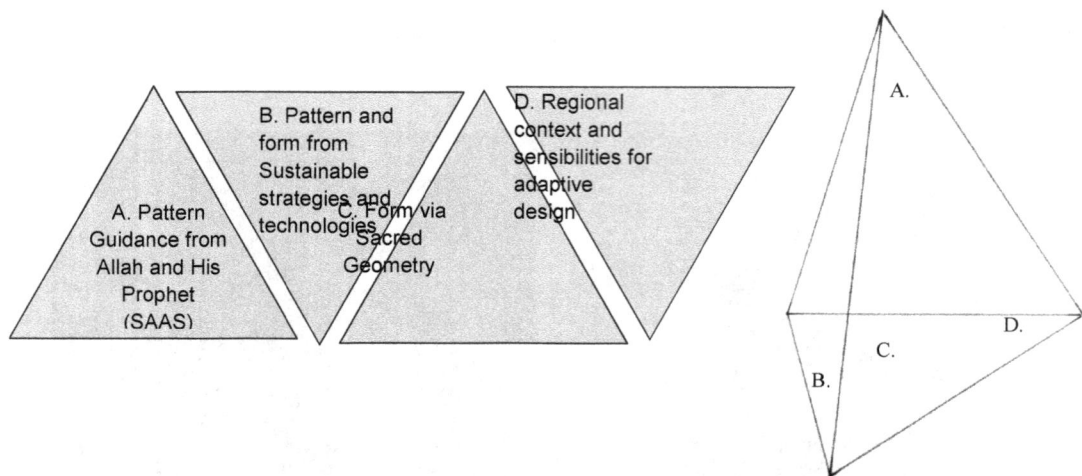

Figure 3. Elements of Architectural Design Guidance in Islamic Architecture (Source: Author).

This guidance includes the following design functions:

*   Protecting the privacy of inhabitants through window placement, use of bent entrances and window screens
*   Encouraging pedestrian movement through the community
*   Including the community building typologies of mosques, homes, educational facilities, and markets within easy access of each other
*   Building in waste and water recycling facilities, including composting
*   Conserving energy through the use of local building materials, building orientation, and building design, making use of natural site features.
*   Using geometry to shape the design to maintain a human scale
*   Incorporating cultural norms and perceptions of beauty

By detailing adaptive Islamic principles, this research aims to facilitate more discussions in the architectural design community of all faiths and provide guidance for the creation of oases of order and peace among the semantic disorder of our time. The absence of projects utilizing these principles has led to the antiquation of Islamic architecture, when in fact, as the research confirms, it has a vital role to play in future development. Projects based on these principles provide a means for Islamic communities to develop built environments that reinforce and strengthen their faith while creating living infrastructure that can adapt to future generations.

ACKNOWLEDGEMENTS
This paper is an elaboration on thesis research towards an MSc in Architecture and Urban Planning in Muslim Societies at Qatar Faculty of Islamic Studies completed in Spring 2014.

REFERENCES

Abdel Haleem, H. (Ed.) (1999). *Islam and the environment*. London, U. K.: Ta-Ha Publishers.

Abu Dawood (2008). Sunan Abu Dawood, 5 Volumes. In Ali Za'I, H. (ed.); Qadhi , Y. (trans.). Riyadh: Darussalam Publishers & Distributors.

Adrian Smith + Gordon Gill Architecture Designs Great City, a Sustainable Satellite City to Begin Construction This Year in Chengdu, China" (2012, October 12) (Press Release) Adrian Smith + Gordon Gill Architecture. Retrieved from http://smithgill.com/news/great_city_press_release/

Akbar, J. (1988). *Crisis in the built environment: The case of the Muslim city*. Leiden: E.J. Brill Publishing Company.

Al-Bukhari, M. (1997). *Sahih al-Bukhari*, 9 volumes. M. Muhsin Khan (trans.). Riyadh, Saudi Arabia: Darrusalam Publishers.

Al-Muwatta of Imam Malik (1982) *Al-Muwatta of Imam Malik*. A. at-Tarjumana and Y. Johnson (trans.). Norwich: The Diwan Press.

Al-Qur'an (n.d.). Yusuf Ali (trans.) Madinah, Saudi Arabia: King Fahd Holy Al-Qur'an Printing Complex.

Al- Sakhawi (n.d.). *Maqasid al-hasana fi Bayan Kathir min al-Ahadith al-Mushtaharat 'ala al-Alsinah*. Dar Al Kotob Al-Ilmiyyah.

Al Tirmidhi (n.d.). *Sunan al-Tirmidhī*. In A. Al-Tāzī (Ed.). Cairo, Egypt: ADD.

Ardalan, N. et al (1973). *The sense of unity: The Sufi tradition in Persian architecture*. Chicago, U. S.: University of Chicago Press.

Arkoun, M. (2002). Spirituality in architecture. In A. Petruccioli et al (Eds.), *Understanding Islamic architecture* (p. 3-8). London, U. K.: Routledge Curzon.

Ashghal (2013). *Doha north sewage treatment works*. Retrieved from http://www.ashghal.gov.qa/en/Projects/Pages/ProjectDetails.aspx?pid=231&affid=3&Category=Drainage+Projects

Benkari, N. et al (2011). Architectural education in United Arab Emirates. *AHU Journal of Engineering & Applied Sciences, 4*(1).

Bianca, S. (1994). *Urban form in the Arab world: Past and present*. London, U. K.: Thames & Hudson.

El-Wakil, A. W. (1981a, April). Being there: Interview with Abdel Wahed El-Wakil. *Domus, 616*, p 18-21.

El-Wakil, A. W. (1981b). Hamdy house, Cairo. *MIMAR architecture in development, 1*, p. 58-61.

Fathy, H. (1986). *Natural energy and vernacular architecture: Principles and examples with reference to hot arid climates*. Chicago, U. S: The University of Chicago.

Fazlul, K. et al (1999). *Islam and the environment*. London, U. K.: Ta-Ha Publishers.

Foltz, R. C. et al. (2003). *Islam and ecology: A bestowed trust*. Cambridge, M. A.: Harvard University Press.

Frings, M. (2002, Winter). The golden section in architectural theory. *Nexus Network Journal, 4*(1), p. 9-32.

Galdieri, E. (2002). Project and tradition. In A. Petruccioli et al (Eds.), *Understanding Islamic architecture*. London, U. K.: Routledge Curzon.

Green Cities California (2013). *Green cities California*. Retrieved from: http://greencitiescalifornia.org/index.html

Hakim, B. S. (2006). *Arabic Islamic cities: Building and planning principles*. London, U. K.: Kegan Paul International.

Ibn Adi et al. (1938). *Al Hadis: An English translation and commentary of Mishkat ul Masabi (Vols I and II)*. In F. Karim (Ed.). Calcutta, Pakistan.

Ibn Habban (1949). *Rawdat al-'uqala' wa-nuzhat al fudala*. Cairo, Egypt.

Ibn Majah, M. (2007). *Sunan Ibn Majah (5 Volumes)*. Riyadh, Saudi Arabia: Darussalam Publications.

Izzi Dien, M. (1997). *The environmental dimensions of Islam*. Cambridge, U. K.: Lutterworth Press.

Khan, H. U. (1982). Traditional housing design in the Islamic countries of Asia. In M. B. Sevcenko (Ed.), *Urban housing*. Cambridge, M. A.: Aga Khan Program for Islamic Architecture.

Lawlor, R. (1982). *Sacred geometry: Philosophy and practice*. London, U. K.: Thames & Hudson Ltd.

Ludwig, A. (2006). *Create an oasis With greywater (5th Edition)*. Santa Barbara, U. S.: Oasis Design.

Making Sand Lime Brick (2012). *Heidelberger kalksandstein*. Retrieved from http://www.heidelberger-kalksandstein.com/heidelberg-sl-brick/making-sand-lime-brick/natural-building-materials.html

Muslim (1956). *Sahih Muslim (5 volumes)*. In M. 'Abd al-Baqi (Ed.). Cairo, Egypt.

National Association of City Transportation Workers (2016). *Streets: Urban street design guide*. Retrieved from http://nacto.org/publication/urban-street-design-guide/streets/.

One Green Home (2013). *One green home*. Retrieved from http://onegreenhome.org/what-are-solar-panels/

Private Engineering Office (n.d.). *Private engineering office*. Retrieved from https://www.linkedin.com/company/private-engineering-office

Pew Research Center (n.d.). The future of the global Muslim population. Retrieved from http://www.pewforum.org/files/2011/01/FutureGlobalMuslimPopulation-WebPDF-Feb10.pdf.

Marçais, W. (1928). L'islamisme et la vie urbaine. *Comptes rendus des séances de l'Académie des Inscriptions et Belles-Lettres, 72e année*(1), p. 86-100.

Nasr, S. H. (1968). *Science and civilization in Islam*. Boston, M. A.: Harvard University Press.

Pope, A. U. (1965). *Persian architecture*. London, U. K.: Thames & Hudson.

Salingaros, N. (2014). *Unified architectural theory: Form, language, complexity*. Levellers Press. Retrieved from http://www.archdaily.com/tag/unified-architectural-theory/.

Salingaros, N. (2006). *A theory of architecture*. Solingen: Umbau-Verlag.

Serageldin, I, & Shluger, E. (2001). Introduction. *Historic cities and sacred sites: Cultural roots for urban futures*. Washington, D. C.. p. xi-xix.

Smith, M. (1944). *Al-Ghazzali the mystic*. London, U. K.: Luzac & Co.

Weather Underground (2013). *Weather history for Doha, Qatar*. Retrieved from http://www.wunderground.com/history/airport/OTBD/2011/1/2/CustomHistory.html?dayend=2&monthend=1&yearend=2012&req_city=NA&req_state=NA&req_statename=NA

AUTHOR

Sabir Nu'Man
Ashghal Public Works Authority,
Doha, Qatar
Email: snuman1978@hotmail.com

6

COPING WITH CROWDING IN HIGH-DENSITY KAMPUNG HOUSING OF JAKARTA

Evawani Ellisa
Universitas Indonesia,
Kampus UI Depok, 16424 Indonesia

Email address: ellisa@eng.ui.ac.id

Abstract
This study aims to draw attention to the architecture of kampung housing, as an attempt to identify those circumstances under which people live in the context of limited space. A kampung housing is a dense non-formally planned cluster of residential dwellings in urban area, which are packed together in a contiguous area created by a large number of migrants. We tried to determine the way in which the spaces are arranged into a place to live, which implies a certain dynamic of survivability among the kampung's inhabitants. The research methodology is conducted with questionnaire surveys, interviews, and detailed observations of daily life cycles, dwelling elements, and the pattern of domestic space arrangements. The study revealed that the characteristics of particular high-density settings have been adapted so that kampung inhabitants devised a particular set of rules and behavioral strategies to cope and support themselves in crowded situations.

Keywords: *Crowding; Dwelling; Space Arrangement; High Density Kampung; Behavioral Strategies*

INTRODUCTION

This article is an outcome of the Research Project Megacity and Global Environment funded by Research Institute for Humanity and Nature (RIHN), Kyoto in collaboration with Universitas Indonesia. It aimed to identify the city of Jakarta as being "a latecomer megacity", i.e. fast growing, but without long-established urban patterns and efficient infrastructure. Jakarta's population reached 9.99 million in 2015 and it is estimated the population will reach around 11.50 million in 2035 (BPS, 2015). The ever-growing mass migration of people descending upon Jakarta has intensified the problem of formal housing provision. The urban lower class inhabitants have to live in the fast-growing, self-emerging, and densely populated settlement called *kampung*s. This hardship of *kampung*'s living has been intensified by crowding; either inside the house itself or the houses is crowded onto almost all available space. How do people cope with crowding and survive living in such densely populated place?

Crowding generally refers to people's psychological response to density; that is, to their feeling of being crowded, having a lack of privacy or an increase in unwanted interactions or psychological distress (Greenfield et al., 1973; Gove at all., 1979; Crothers et al., 1993; Jazwinski, 1998). Research on the consequences of crowding in urban housings has been done on a massive scale by urban sociologists and psychologists. They argue that crowded housings condition leads to poor mental and physical health, poor social relations at home, and have negative effects to child care (Baldassare, 1988; Freedman, 1970; Lawrence, 1974; Rodgers, 1981; Stokols, 1972). Yet, according to Altman (1978), crowding is a psychological state, a subjective experience that refers to a feeling of having very little space. Hall (1966) emphasized the theory that crowding is a subjective experience which appears to significantly vary across different cultures. Yet most of the scientific, cross cultural studies on crowding have been conducted within North America (Evans et al., 2000).

As crowding exists in three different modes of situational, emotional and behavioral perspective, it is a researcher's task to identify the personal, social and physical factors that lead individuals to label and experience crowding (Gifford, 2002). This paper does not intend to focus on the consequences of crowding nor measure prevalence factors associated with crowding. It attempts to reveal what the residents are doing and determine not only at what point they are in a crowded situation but also the ways for relieving crowding. The research question is: How do the *kampung* dwellers develop strategy to cope with crowding? As *kampung* inhabitants have clearly adapted to density, how does crowding affect their method of spatial arrangements to create decent domestic living space?

THE AIMS OF THE STUDY

Many studies have revealed that informal unplanned areas are densely crowded and uncomfortable, yet most studies are not accompanied by detailed study and a complete recording of the houses. Much study is focused on the effect of housing on health (Ambrose et al. 1996b; Baker et al., 2000; Muller et al., 2007; Evans et al., 2003) and the neighborhood (Brown et al, 2009; Curry et al, 2008; Sampson, 2003; Schulz et al, 2000). However, the relationship between crowding and health is extremely complex and is influenced by a number of covalent variables included the condition of housing and socioeconomic factors, such as income, employment status, and education (Fuller, 1993).

In terms of housing, although *kampung*s are not always equated with slums (Mc Carthy, 2003), many researches have concentrated on *kampung*s as being the urban underclass habitation with a poor state of infrastructure and degraded environmental condition (Silas, 2010; Jellinek, 1995; Suparlan, 1984; Nas, 2000; Somantri, 2007; Silver, 2011). Some researchers have focused on *kampung*s within the context of the history of Jakarta (Colombijn, 2010; Abeyasekere, 1987), economic and social life (Tunas, 2011), culture (Funo et al, 2005) and the transformation process of *kampung* (Funo et al, 2002; Funo et al, 2004). There is only one study on *kampung*s and crowding that focused on the issue of cultural differences in perceptions of crowding through identification of a wide range of expression related to crowding (Clauson-Kaas et.al, 1997). This paper addressed the lack of research on *kampung*s that is predominantly concerned with spatial constraint caused by crowding and the motivation to eliminate or reduce their salience. The study also enriches debates on the wide spectrum exploration and investigation on housing issues, especially in South East Asia (Salama, 2015).

As people living in *kampung*s are disregarded as subaltern or as "others", there appears to be little research that focuses on their living environment and everyday life. De Certeau in his famous conceptualization of strategies and tactics enhances the significance and necessity of analyzing "ordinary people's daily lives" (De Certeau, 1984). To do this, he emphasizes on people's practices as the multiple domains that demonstrate societal data. He states that everyday practices are not random actions of people; on the contrary, they have logic to be understood. Through analyses of power mechanism on daily life practice of "the others", he declared that ordinary people, in fact do not surrender to power and its regulations, but form invisible resistance.

As a preliminary attempt to recognize real-life circumstances under which people live in high density living environments, this research tried to determine the way in which overcrowded houses in the *kampung* are used and adapted as places to live. Bourdieu mentioned that most significant micro practices conceive the main element of whole society should be the "house". Dwellings are places that create different forms of discourse, lifestyles and practices by reproductions (Bourdieu, 1996). De Carteau stated that: "[t]his fragment of society and analyses is first of all the dwelling, which is as we know the reference of every metaphor. Through the practices that articulate its interior space, it inverts the strategies of public space and silently organizes the language (a vocabulary, proverbs, etc.)" (Certeau, 1984). Depending upon the theoretical framework on an individual tactic in daily life, this research tried to bring everyday life and mundane practices to the forefront. The main focus of this study was *kampung* dwellers who were active agents of their own daily life practices. Within this existence, as "different ones", they can produce new forms of resistance against generally accepted ways of living in formal housing.

METHODOLOGY

This study focused on *Kampung* Cikini at the district level or the Kelurahan Pegangsaan Central Jakarta. The case study is a representative of typical high density informal settlement in Jakarta. According to data from the Kelurahan, in 2014 *Kampung* Cikini covers 1.5 hectares and has a population of 3.784 inhabitants (2,522 person/ha). Data gathering consisted of two phases. The first phase was a questionnaire and survey conducted by the author in collaboration with Life Style Team of Megacity and Global Environment Research Project (Kato, 2012). The team distributed structured and semi-structured questions to 146 respondents in *Kampung* Cikini in September, 2012. All respondents were selected by a principle of volunteer participation. The investigation focused on how inhabitant' lifestyle impacted their behavior through questions on daily activities such as work, leisure, house repair activities, and social cultural activities.

The second phase aimed to find out through case studies how dwellings at *Kampung* Cikini had been arranged to cope with crowding. Interviewers went door-to-door of randomly selected blocks of *kampung* dwellings to observe the physical condition of the houses and seek households who were willing to be selected as survey participants. As Flick (1998) mentioned, in qualitative research, it is the relevance to the research topic rather than the representativeness which determines the way in which respondents/objects are selected. Eventually, only 20 respondents were willing to be selected as subjects for case studies. However, the number was sufficient to provide the relevant information for the research. The selected houses sizes ranged from 6 to 42 m2. These houses typically represented the range from the smallest and the biggest houses. Each case categorically represented the house's type based on layout and modification.

To describe the way in which *kampung* dwellers arrange their domestic space, students visited selected houses for semi-structured interviews and detailed mappings of their dwelling situations based on precise measurements, photographs, and video documentation. Groups of 3 (three) students intensively worked on each house. The first student conducted the interview with house owners in their mother tongue; the second made interior sketches; and the third assessed other aspects of the house using photographs and videos. The students drew the use of space as well as the fixed and unfixed furniture used in the house. For each category, the students had to spend time to develop a rapport with the respondents to allow detailed investigation of their private domestic spaces and prevent inaccurate information. To reveal the spatial representation, students encouraged the inhabitants to talk not only on how they utilized the space, but also how they appropriated spaces as their home and what that spaces meant for them.

In doing the analyses, the spatial representation of recorded data was shifted into deeper architectural spatial knowledge within the richness of spatial usage. This knowledge can only come from residents, although they might not be able to articulate space as an everyday knowledge. Interpretations were developed from what the space represented and the sense of what respondents did not say about the space. As Miller (1987) pointed out, the conscious reading of space is often revealed in nonverbal clues. For that reason the residents' own voice and narratives were not always employed as quotations. From twenty cases studies investigated during research, based on categories on how they negotiate with crowding, finally seven cases were selected to be presented in this paper.

HISTORY AND ACTUAL CONDITION OF KAMPUNG CIKINI

Originally meaning "villages", the irregularly formed *kampung*s had already existed for a long time and constituted one of the typical features of towns and cities in Indonesia (Rutz, 1987). During the Dutch colonial era, the formal urban development plan bypassed the existing low density *kampung*s to integrate them into urban areas, but without ample provision for urban utilities and facilities. As a result, these settlements tend to be evenly scattered throughout the formally planned built-up areas. The *kampung*s provided dwelling places for the laborers at the wharves, warehouses; industries and public works as well supplied domestic helpers for middle class neighborhoods (Castels, 1967). By the middle of the 20th century, a massive numbers of migrants migrated to cities and found *kampung*s as attractive, easily accessible and cheap locations to live. New ad-hoc buildings soon infiltrated the vacant lands and transformed low density *kampung*s into high density, heterogeneous and sub-standard settlements. In 1969, almost 75% of the total population in Jakarta lived in *kampung*s (Rachman, 1995). They were the urban poor who had to face severe social and economic problems that forced the local government to declare Jakarta as closed to migration in 1970. In 1974, the government enacted *Kampung* Improvement Program (KIP) to improve the existing housing stock and allow for the provision of service. Nowadays, while it is very difficult to obtain the exact size of the population living in the *kampung*s due to the complexity of data, the Urban Poor Consortium (UPC) Jakarta reported that 20-25% of the total population in Jakarta lived in *kampung*s with an additional 4-5% living illegally along riverbanks, an empty lots and in flood plains (Urban Poor Consortium www.upc.org, 2010).

The history of *Kampung* Cikini is traced back in early 20th century, when the Dutch developed the district for the Dutch elite class in Menteng Estate. Native inhabitants lived along the bank of the Ciliwung River soon took the opportunity to provide service for upcoming well-to-do Dutch inhabitants. At the same time, the Dutch also developed the railway infrastructure through *Kampung* Cikini. In the 1960's, when the government closed down railway in the former location of *Kampung* Cikini, the railway track was overlaid with asphalt to transform it into a vehicular access. The railway yards and embankments were left vacant. People took advantage to squat and then built shelters in the empty spaces that they became denser every year.

While those who squatted the ex-railway yard claimed to have permission from National Railway Company (PT Kereta Api Indonesia), other inhabitants have declared semi-legal land ownership rights granted from the colonial authority called *hak girik* (tribal land right). This is a kind of land certification that is acknowledged by government, but it does not entitle them the rightful ownership of property (*hak milik*). Throughout the years, *kampung* householders have acquired a sense of complacency as they have received benefits such as electricity and paying *Pajak Bumi Bangunan* (PBB) or taxes for building and land, thus implying their land use rights. *Kampung* Cikini is located in one of the most expensive and sought after sites in Jakarta with close proximity to Cikini Train Station, the flower market (Pasar Kembang) a six-storey shopping center. The surrounding developments, such as hospital, hotels, offices, restaurant, and so forth offered an economic opportunity for low-income dwellers of *Kampung* Cikini.

Kampung Cikini is designated as an RW (*Rukun Warga* or Community Unit Group) and is divided into 13 RT (*Rukun Tetangga* or Neighborhood Unit Group), of which each RT includes 50-60 households. The demographic profiles of the inhabitants are rather diverse; they are coming from many different ethnic groups from around the Indonesian archipelago. The unclear status of land does not necessarily discourage the practice of an informal land market where many houses in the *kampung* were passed down to different hands across the years. Owing to the great location and accessibility of employment, recently the demographic texture of *Kampung* Cikini tends to be diversified by formal sector workers and educated people. The strategic location of *Kampung* Cikini and the rises in land value attract newly middle income occupants who seeking the immediacy of home in the city center.

As in most *kampung*s, social relations among inhabitants are relatively strong. Most of residents socialize and interact with each other at daily basis. As formal regulations were absent, the numbers of social groups and networks helped to establish order in *kampung*. Furthermore, the mandatory in Islamic social-religious customs, which are embraced by 97% of respondents controlled the norms of daily life and bounded the community together. There were 5 (five) mosques in *Kampung* Cikini that indicates the significance of religious institutions to community living.

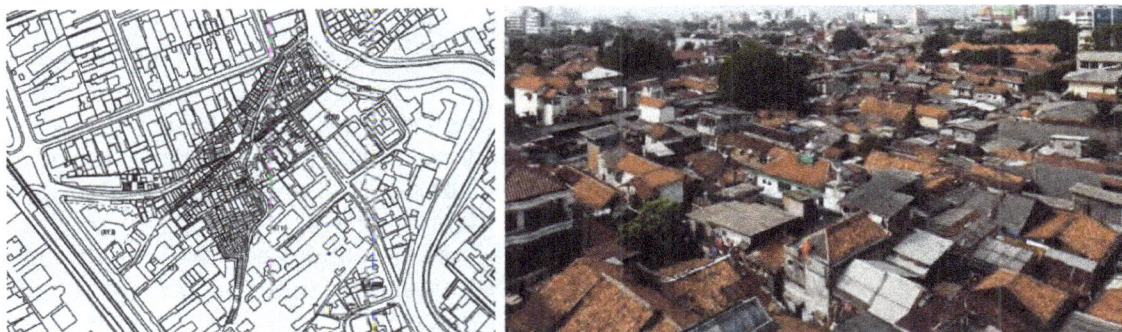

Figure 1. *Kampung* Cikini (Source: Ellisa, 2012)

DWELLING QUALITIES, LIFE STYLE AND NEIGHBORHOOD CHARACTERISTIC

Surveys and questionnaires about houses in *kampung* Cikini revealed that the building construction approach was typical self-help without any particular design method and refreshingly free from spatial constraints. Buildings are incrementally and gradually developed around the original starter unit or the basic shelter. The results were mostly a blend of old and new structures. Some buildings are remnants of older periods, while others have been recently renovated or completely rebuilt. Residents often add building extensions that occur not only because of the direct pressure from family growth, but also when the family acquired some source of additional funding. The building materials and finish quality are comparable to formal low-cost housing: 87.9% of the respondents used a reinforced concrete post and beam skeletal system with red brick infill. The roof is structurally formed of wood beams with fiber cement corrugated roofing. The ceiling height is typically low, and it does not allow for good air circulation. Flooring is largely ceramic tile over concrete floor.

Figure 2. Some Examples of Houses in *Kampung* Cikini (Source: Ellisa, 2012)

The houses have been expanded in such a way that residents found difficulties to make ample windows and openings to ensure cross-ventilation. As it was essential to withstand the hot humid tropical climate of Jakarta, all respondents had at least one electric fan to control the thermal comfort inside their houses. The assumption that low income, high density housing constrains the possession of home electrical equipment is completely misleading. The questionnaire survey revealed that various electrical appliances were fully used for resident's daily-life activities. Mobile telephones were not necessarily seen as a luxury or even as a status symbol, but these devices provide a sense of withdrawal from the main residential space for people who had no private space of their own. High density living does not constrain those who have hobbies and interests that occupy comparatively large spaces, such as keeping birds and breeding fish.

For the inhabitants, the *kampung* represents more than just a space to live; it offers security and stability. Kent (1990) pointed out that the house cannot be seen in isolation from the settlement, but it must be viewed as part of a total social and spatial system that relates to the way of life of a particular settlement. Unlike vertical urban tenements, the morphological arrangement of the *kampung* offers some advantages. As there are no formal rules and or rigid distinctions between private and non-private space, it is very normal for residents to make use of the space immediately outside their homes to do their laundry, keep belongings, cook, prepare food, and eat meals. Living together as a community the inhabitants secure about leaving their doors and windows open. High visibility inside and out gives neighbors plenty of opportunities to interact and chat. Since everybody knew everyone else's business, there is no need to interfere and the residents could live together in a spirit of mutual reliance.

Figure 3. Outdoor Alleys of Gangs in the *kampung* (Source: Ellisa, 2012)

QUANTIFIABLE MEASURE OF THE HOUSE

Before understanding how residents deal with crowding, we need to measure the quantifiable manner of dwelling units in *kampung* Cikini. A questionnaire survey of 146 respondents revealed that 24.3% of the respondents were living in households with 4 (four) individuals, 23.4% were living with 3 (three) individuals, 16.8% were living with 5 (five) individuals, 12.1% were living with 2 (two) individuals, 8.4% were living with 1 (one) individual, and 7.5% were living with more than 6 (six) individuals. The average number of individuals in the household was 3.93. In comparison with the situation during the urbanization boom in previous years, nowadays those who live in the *kampung* are not predominantly members of an extended family.

Regarding the number of rooms they lived in, it revealed that 39.3% of the respondents lived in a household with 1 (one) room, followed by 32.7% who were living in 2 (two) rooms, 19.6% were living in a household with 3 (three) rooms, 5.6% were living in a household with 4 (four) rooms, and 1.8% living in a household with over 4 (four) rooms. The median of room number of rooms was 1.5rooms/household. When compared with the average number of individuals in the household and median number of rooms, each person in a household had 0.38 rooms/person. This indicated that the dwelling unit at *Kampung* Cikini was far below the range 0.6 rooms/person to 0.8 rooms/person of the European standard (Edwards et.al., 1994).

To understand the size of habitable floor space, we categorized the houses ranging within an interval of 10m2. The survey revealed that 22.4% of the houses had room sizes in the range of 11–20 m2, 23.4% had room sizes in the range of 21–30m2, 7.5% had room sizes in the range of 31–40m2, 14.0% had room sizes in the range of 41–50m2, and the rest were found to be distributed randomly between 50 and 100m2. The median floor area was 23.5m2. A comparison between the median of floor area and the average number of individuals, with the average actual room consumption per person showed a value of 5.98m2/person, indicating that *kampung* dwellers have inadequate living space, when compare to Indonesian National Standard of 9m2/person. It cleared that the quantifiable manner of houses in combination with the qualities of houses in the *kampung* indicated that *kampung* residents not only face the problem of sub-standard quality of housing but also overcrowding.

DWELLING SPACE ARRANGEMENT

As all buildings in the *kampung* were typical self-help and incrementally developed at the various extended time, there were no two houses the same. Yet, from twenty case studes investigated during the research period, they eventually were classified into 7 (seven) different types based on a combination of the sizes, the characteristics and the spatial arrangements. The selected cases consisted of two extended families, four single families, and two doubled households. Five selected houses were two-storey, one house was a three-storey, and one house was a one-storey. Basically, they were arranged to allocate furniture and user's activities which included access and movement inside the house, or what Bollnow identifies as the space of action (Bollnow, 2011). Based on how the inhabitants negotiate with crowding, all cases were clustered into four categories as follows: 1) a basic shelter; 2) a common house; 3) a house as an income *generator*; and 4) an ideal house.

Basic shelter type

Basic shelter was a very small house that it was almost impossible to include a wet area inside the house. As a consequence of crowding, the wet area in the *kampung* is defined as the space for a kitchen and a bathroom with or without a toilet. Many older houses that were built when the *kampung* was less crowded might have private toilets. However, as the *kampung* grew getting more and more densely packed, it becomes almost impossible to locate septic tanks. There was a rule that the newly build houses strictly not allowed to have an individual indoor toilet. For that reason, residents had to use the public toilet or *Mandi Cuci Kakus* (MCK) for bathing, washing, and toilet. Since they were sufficiently provided nearby, residents did not find it was inconvenient not to have private toilets.

Sundari House

Sundari's house was a one room shelter, 1.8m in width and 3.4m in length (6.12m2). As Sundari lived with her husband and three children, she needed to take the full advantage of the available space. Here was the place for all daily activities, ranging from sleeping, eating, raising children, studying, watching TV, and cooking. There was hardly any furniture inside the house except storage unit stacked next to the wall to keep all family belongings.

This family could not resist the problem of overlapping activities in their daily domestic life. All domestic activities were carried on with minimum equipment in whenever spaces were left free of belongings. Privacy was totally ignored because there was no compartmentalized space inside the house. Sundari said that intimate relations could only be done when all the children fell asleep. Anticipating crowding, her husband and children spent almost all of their daily activities outside, except sleeping at night. This allowed Sundari and her 2-year-old child to enjoy a more spacious feeling while being at home during the day. Our observation of her house revealed that there was no space to avoid feeling "cramped". Yet Sundari felt, as her family gradually grew and adapted to the space they eventually knew very well how to dwell in it. She said that although the house was barely furnished, it was not bare and empty like a prison cell. Excessive smallness did not mean an unsettling situation as each family member was able to stretch out at one's ease. She said that the important task of the house was to provide a refuge from the outside world. For Sundari's family, no matter how small the house, it fulfilled the basic concept that the dwelling space must give an impression of seclusion.

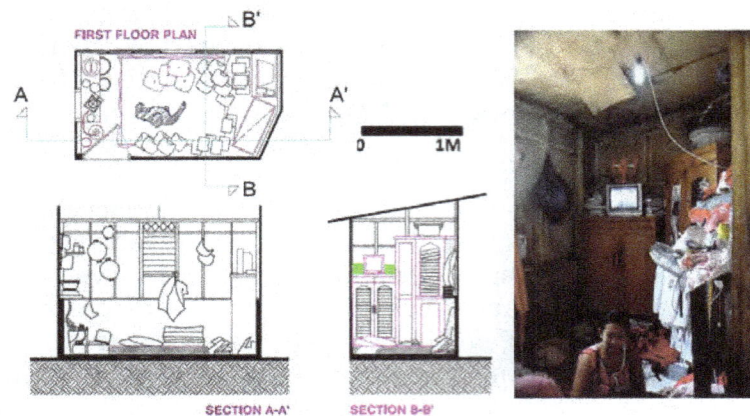

Figure 4. Sundari House (Source: Listiyanti, 2012)

Suheni and Mimin House

The case of non-related, doubled up household was very common in the *kampung*. The second case was this type of house occupied by two extended families that were living under one roof, one on the first floor and the other on the second floor. Both families lived with the elderly. The house was two-storey, each floor size was 2.43m×6.0m (14.58m2). The family living on the first floor was Suheni's family, which consisted of a couple, a paralyzed grandmother, and one child. The second floor was Mimin's family consisted of a couple, grandmother, and two children.

Suheni arranged her living place based on a relative degree of privacy, from back to front. She put the bedroom for herself and her husband in the back of the house, a divided room for her daughter and grandmother in the middle, and a multi-function room at the front. There were stairs in the middle of the room as the only access to Mimin's family to their living place upstairs. Two stoves-tops were located close to the stairs, one owned by Saheni and the other owned by Mimin. However, both housewives rarely cooked and preferred to buy edible food outside. In the multi-functional room, Suheni put all of her family belongings. As there was a large amount of furniture and belongings, the space for movement inside the house was very poor.

Like Sundari, when asking about her feelings in such an overcrowded house, Suheni demonstrated an impression of privacy in her house. Suheni and her family spend most of the time at home whenever possible. Crowding was a factor that depreciated the spatial quality of her home, but it did not appear to be an overriding deterrent to happiness. The smallness of her house did not restrict her family in spending most of their time at home. As Suheni did not enjoy socializing with her neighbors, during leisure time she preferred to watch TV while accompanying her daughter when she was doing her homework. She said disorder and clutter neither have a disquieting effect, nor disturbing her daily activities at home.

The second floor was occupied by Mimin's family. The family comprised a 70-year-old grandmother and a couple with a 7-year-old son and a baby. Here, the conditions were worse. There were two rooms, one was for the grandmother's bedroom and the other was the parent's room with their children.

Except for sleeping, the dwelling space was inadequate to provide sufficient internal space for basic daily living activities. During our several visits, we found that Mimin's family coped with this severe crowding by spending their daily activities outside as much as possible. Crowding forced this family to deal with two constraints. First, their shelter did not have any connection with the earth's surface. To reach their own house on the second floor, this family had to interfere to other family territories at the first floor. Second, there was scarcely any space to escape inside the house, as the whole space was packed with belongings.

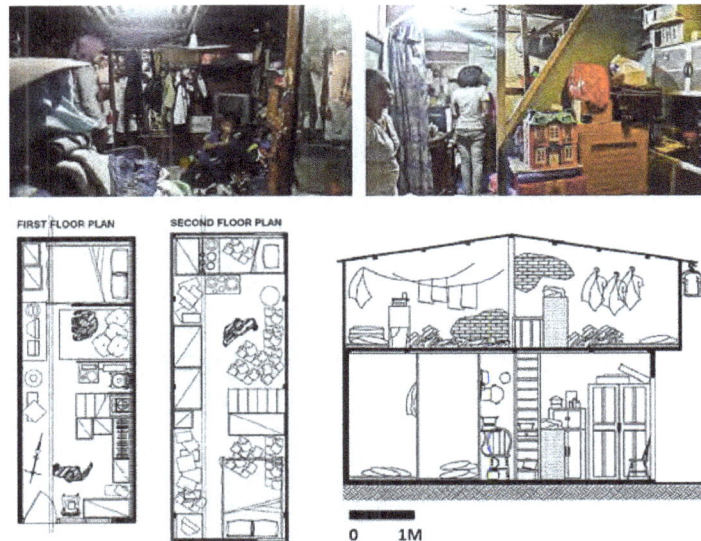

Figure 5. Suheni and Mimin House (Source: Listiyanti, 2012)

It was not surprising that Mimir did not express any feeling of intimacy with her home. She realized that her only choice was to stay outside most of the time. Normally, one might consider a house as the refuge from outside. However, for Mimin' family the house was barely more than a storage space. To escape from overcrowding, the exterior of the houses or "the outside world" was the best place. As the organic pattern of dwellings with *kampung* created no defined boundaries between private and public spaces, Mimin's family extend their territories beyond their actual dwelling and considered the public space simply as part of their own territory.

Common House type
The two houses below were examples of common houses in the *kampung* in terms of the total amount of space, the way in which space had been divided, the availability of space for supporting daily living activities, as well as the availability of a wet area.

Nanan House
The first common house type case was a three-storey house (4.3m×2.75m at the lower levels and 4.3m×3.5m at the upper level), owned by Nanan, who lived with his wife and adult son. As the head of the neighborhood community (*Rukun Tetangga*), Nanan often had visitors and discussed community matters with his neighbors. Therefore, he needed to set up the first floor almost entirely as public space. He located the wet area on the first floor, consisting of a mini-kitchen, washing machine and a bathroom without a toilet. Like most of the other residents, Nanan family would use the communal toilet at the MCK for solid waste, but they preferred to urinate on the cement floor in the bathroom. With no proper plumbing inside the house, the waste water simply flowed down into the gutter outside the house.

On the second floor, Nanan divided the area into two parts, one was a space for watching TV and family gathering, and the other was the space for the "master" bedroom. On the third floor, the space was divided into two parts, one was an area for drying clothes, ironing, and storage, and the other was for his married daughter. Yet the room was often empty, as his daughter eventually preferred to live in another place with her husband.

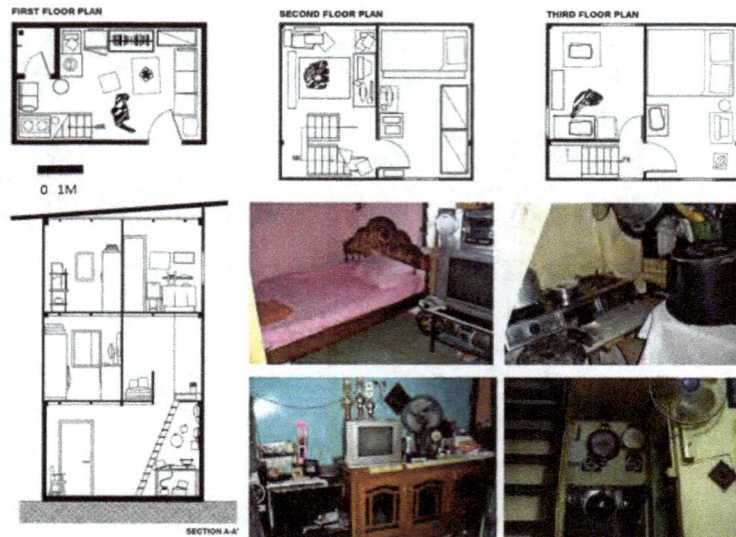

Figure 6. Nanan House (Source: Listiyanti, 2012)

The Nanan's house provided the basic internal functionality that included space for furniture needed by the residents (including occasional visitors), space to access the furniture, space to move around the house, space to undertake living activities (washing, dressing, cooking, eating), and space for the storage of daily items. The separation between the rooms allowed for the required level of privacy, but the size of the house was too small to provide adequate space to avoid the feeling of being "cramped". Nanan expressed the difficulty in deciding whether to fill up or to utilize the space inside his house. In order to function well, inhabitants needed to lessen the sense of individuality and maintain tolerance each other.

Een House

The second common house type case was a two-storied house with a size of 2.45m×8m (19.4m2) at the lower level and 2.45m×3.8m at upper level (9.31m2). Een shared the house with her married daughter who lived with her husband and twin daughters. The first floor was divided into two parts – one was at the front with a multipurpose room that accommodated several simultaneous activities. During the day it functioned as a common space and in the night it was transformed into a sleeping space for Een and her grandchildren. In the back of the house, the family utilized an extended narrow longitudinal space for a bathroom with a toilet and a kitchen.The second floor had become a multipurpose space, irregularly arranged as the area for watching TV, storage, and drying clothes. Een arranged a bedroom for her married daughter by putting a simple divider for privacy.

The Een's house represented a common house type in the *kampung* as we often noticed similar ones during our field observation. These houses were arranged with no separation between the spaces for storage and living activities. Nevertheless, the house showed a strong flexibility and adaptability for multi-family accommodation. On one hand, there was no effort to arrange the house into a more comfortable place to live, but on the other hand, it implied the effort of multi-family members to adjust their lifestyles to live in crowded conditions under the same roof. For family members, this was not a problem as they were living with family members.

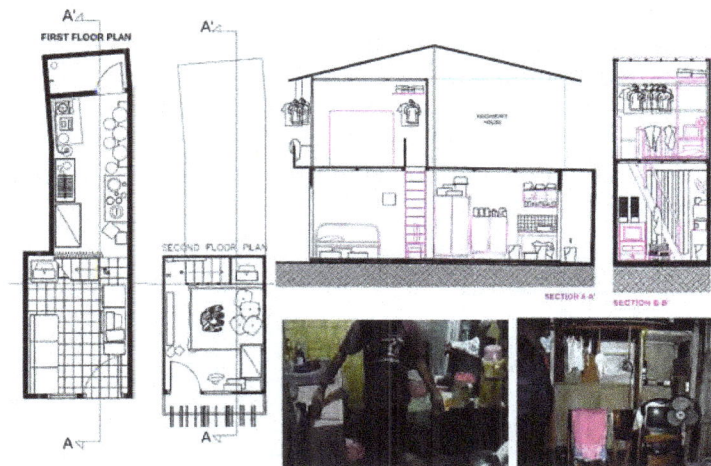

Figure 7. Een House (Source: Listiyanti, 2012)

During the interview, Een family members expressed little interest in improving the use of space. It was not that they were dissatisfied or felt they already had sufficient space, but rather because they were not sufficiently motivated. Ironically, this reluctance to improve the spatial arrangement of the house seemed to be dominantly represented in the character of domestic space in the *kampung* housing. The two generations also exhibited a marked contrast in consumption attitudes that triggered overcrowding, as all belongings of this family were not always essential items. Een's daughter mentioned the generation gap between her mother and herself when she referred to her mother who continued to use of things her daughter considered either old or useless.

The house as an income generator
Lack of access to the formal job market has encouraged the inhabitants to use their crowded domestic space for generating income. Many houses accommodate these activities such as a *warung* or a stall, services, private rented rooms, and small enterprises production. Each had different characteristics with respect to space organization and the nature of work. The double functioning of the dwelling as both home and workplace had transformed the domestic space into an essential shelter for subsistence activities. The alteration of domestic space into a home base for income generation also indicated the security of tenure of living in the *kampung*. As the dwelling space itself faced the problem of congestion in dealing with domestic activities, the inhabitants have to find their own working space within these limitations. There are two case studies for the house as income generation, the Atin house and the Rosadah house.

Atin House
The first case of the house as an income generator was Atin's house. Atin was a single mother living with her adult son and her married daughter with one child. Owing to its strategic location, Atin had fully utilized the house to generate the family's income through culinary production. As cooking activities needed more space, when compared to Jaya's house, Atin's house was bigger (4.2m×4.7m or 19.74m2 at the lower level and 4.2m×5.9m or 24.78m2 at the upper level).

Atin used the space sparingly between food production and daily living based on time differentiation. During the day, the entire area of the first floor became the kitchen and food stall. She simply aggregated the space required for cooking activities which were butted up against one another without any leftover space. She collected water for cooking from the faucet in the bathroom in the corner of the house. At night, Atin cleared all the cooking appliances, put them near the wall, and transformed the space into her sleeping area, a place for relaxation, and for enjoying TV. She dedicated almost the whole day for cooking while at night what she needed was merely a place for a little relaxation and sleep.

The second floor was entirely used as the living space for her children, consisting of one bedroom for her son, one bedroom for her daughter's family, and one empty bedroom. There was nothing special about the second floor because it represented a common living space and arrangement in the *kampung*.

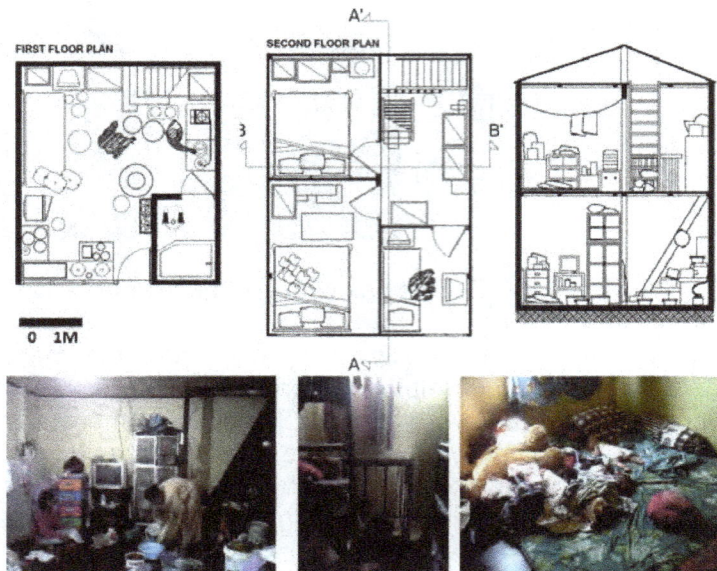

Figure 10. Atin House (Source: Listiyanti, 2012)

Rosadah House

The second case was Rosadah's house. It was an irregular shape (7.3m×2.3m or 16.79m2 at the lower level and 8.1m×3.1m or 25.11m2 at the upper level). Rosadah was 70 years old and lived alone. She occupied the first floor and arranged it in the same way as typical houses in the *kampung*. The house spoke of crowding that was not imposed by the spatial limitations, but produced by owner's own desire. Furniture took up a large amount of space and left almost no space for moving around. There was a small kitchen, but she almost never used it. Rosadah was fully aware, as she was living alone, she would have more space if she cleared out unused stuffs. Yet, she needed to keep so many things related to her past experience or when things were not easy to get.

Three rooms on the second floor used to be occupied by her children. She left one room for her grandchild who often visited her and rented the other two rooms for families. One was a couple without a child and the other was a family with two children. The renters accessed their rooms from the stairs outside and shared the toilet with the homeowner on the first floor.

Rosadah resorted to renting out the empty rooms to tenants for income. Without any spatial adjustment and rearrangement, her house was easily adapted to accommodate the change of family composition. She transformed the house from a single family into a multi-family dwelling. The presence of tenants on the second floor only required minimal interface, although they have to share a toilet. Neither the owner nor the renter considered this situation as a constraint.

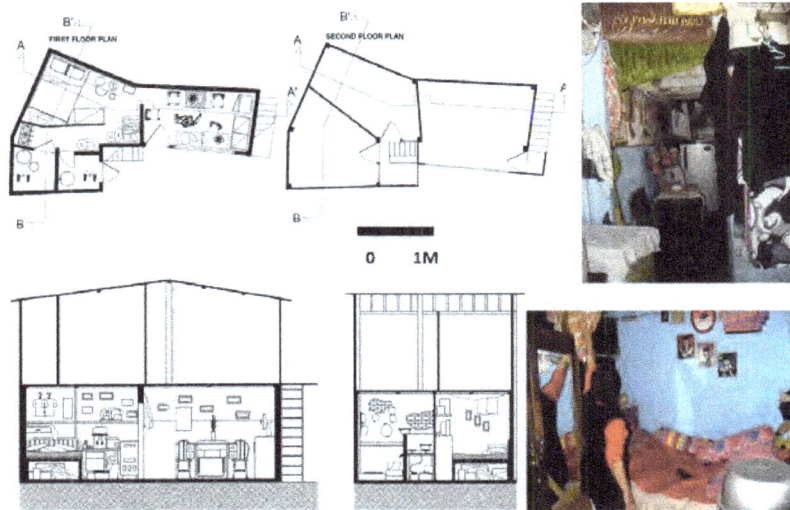

Figure 11. Rosadah House (Source: Listiyanti, 2012)

Ideal House – Siti House

We choose a small house (1.75m×4.38m or 7.66m2 at the lower level and 1.75m×5.18 m or 9.06m2 at the upper level) owned by Siti as the Ideal House. Here the owner skillfully arranged the small shelter into a neatly compact, multi-functional living space. Siti lived with her husband and two children. Taking full advantage of the available space on the first floor, she planned the circulation inside the house in accordance with the door placement. As a result, the room was not cut in two by the traffic flow as in other typical houses.

She clustered items into separated units to eliminate clutter and increased the floor space for circulation. She located a private bath, toilet, and washing area under the stairs, with the stove in the corner. In the kitchen, she utilized the space in-between the table and the ceiling height to store cooking utensils. She carefully arranged all furniture parallel to the walls and categorized them based on the character and closeness of activities. She arranged the space at the corner for watching TV. When the door was closed, the space in front part was transformed into a more spacious place. Siti decided not to hoard things and made a special effort to keep their possessions in minimum.

Figure 12. Ideal House (Source: Listiyanti, 2012)

As in other cases, she also utilized the second floor for sleeping and storage. She put a double bed at one side of the wall and spread a curtain to divide the bed from the storage area. She eliminated less essential furniture to provide more usable space. There were no laundry clutters because she hung the clothes on the balcony. Finally, she kept less-used belongings in the attic. Siti was a rare example of *kampung* residents. She consciously decided not to accumulate things and made a special effort to minimize possessions. With careful decisions to select specific belongings she developed a sense of spatial logic by making the best use of the space available in her house.

DISCUSSION

All cases discussed in this paper show that although individual differences aroused different ways of compromising lifestyle choices within a high-density situation, the characteristic of particular high-density settings motivated the inhabitants to develop a specific set of rules and strategies to help them cope with crowding. (See Table 1).

The lessons learned from these four categories are as follows:

a) There was no minimum size requirement in the basic habitable parts of dwelling in the *kampung*, because inhabitants always found ways to use and negotiate with the existing living space, notwithstanding the number of persons who live within homes and how they relate to each other. The smallest dwelling basically was arranged into a space for sleeping at night and the storage of "dry" belongings. As inhabitants showed mixed responses, either highly tolerant or by withdrawal to minimize the inconvenience of living in such crowding, further insight is needed to answer how this marginal shelter meets the fundamental standard of health, privacy, safety and morality.

b) In all cases, the addition of the wet area (toilet, kitchen, washing) into the dwelling area increased crowding and poor hygienic practices. It caused space reduction, either in the habitable rooms or for the storage space. In terms of cooking activity, there was continuing uncertainty on whether the kitchen space inside the house had to be maintained or at least reduced, because in many cases, the inhabitants preferred to buy edible food outside. A further insight needs to be considered whether an indoor kitchen manifests a practical function or is simply symbolic attribute within the housewife's realm in the *kampung*.

c) The majority of common houses in the *kampung* sufficiently provided the basic internal functionality of the house with a number of mechanisms to cope with crowding. Better than a basic house, these common houses provide at least a minimum degree of privacy through the separation of rooms. However, the size of the house was too small to provide adequate space to avoid a feeling of being "cramped".

d) The effort of the inhabitants in all cases to compartmentalize the space for sleeping, whenever possible indicated that the lack of privacy from an open-plan arrangement was a major issue. The main concern was about sleeping space for adolescent and adult members of the opposite sex, except the husband and wife.

e) The attitude of the residents to keep things does not appear to be restricted by density. In the absence of spacious rooms for storage, there was no way that the items which the family might wish to be put away could be hidden due to a lack of storage units. The spatial appropriation for the placement of belongings was not always a conscious action, but was often based on what was convenient and close at hand.

f) Crowding is very much associated with household size and composition, which reflects not only the affordability issue but also cultural norms. Yet, crowding lessened as children grow up and leave home. Therefore, crowding was temporary, dynamic and will change over time as the age and composition of households change.

g) The primary motivation cf the spatial arrangements for most of the residents who live at houses hosting economic activities focused on the need to accommodate the space for the "business," rather than the need to create a comfortable living space.

h) The case of the Ideal House revealed that, notwithstanding the size, the inhabitant's ability to arrange a small space through zoning daily activities, clustering items, and eliminating less essential furniture was the key in finding the best way to cope with "crowding".

Table 1. Characteristics of Houses of Seven Case Studies (Ellisa, 2012)

	BASIC SHELTER		COMMON HOUSE		INCOME GENERATOR		IDEAL
OWNER	Sundari	Suheni & Mimin	Nanan	Een	Rosadah	Atin	Siti
SIZE (SQUARE METERS)	6.1	20.1	42	28.9	27.2	41.9	16.7
FLOOR NUMBER	1	2	3	2	2	2	2
NUMBER OF THE ROOM	1	5	6	4	4	9	7
NUMBER OF PERSON	7	9	3	5	8	4	5
FAMILY TYPE	Nuclear Family	Doubled up household	Nuclear Family	Extended Family	Owner and Renters	Extended Family	Nuclear Family
WET AREA							
WashingTap	No	No	Yes	Yes	No	Yes	Yes
Toilet	No	No	No	No	No	Yes	Yes
Bath	No	No	Yes	Yes	Yes	Yes	Yes
SEPARATE SLEEPING SPACE	No	No	Yes	Yes	Yes	Yes	Yes
COOKING SPACE	Not available	Available but rarely used	Available but rarely used	Available	Available but rarely used	Available	Available
STORAGE	Mix & untidy	Mix & untidy	Separated & upkeep	Separated & upkeep	Mix & untidy	Mix & untidy	Separated & upkeep
FURNITURE	Basic	Excessive	Adequate	Excessive	Excessive	Adequate	Adequate
CIRCULATION/ MOVEMENT	Very congested	Very congested	Congested	Congested	Congested	Sufficient	Sufficient
DAILY ACITIVITIES	Overlapping	Overlapping	Overlapping	Separating	Overlapping	Partially Overlapping	Separating
DEGREE OF PRIVACY	No privacy at all	No privacy at all	Minimum	Minimum	Minimum	Adequate	Adequate
STRATEGY TO COPE WITH CROWDING	Spend almost all the daily activities outside, except sleeping at night	Suheni: Arrange a relative degree of privacy. Mimin: Extend dwelling territory outside the house.	Arrange the space to provide basic space for furniture, access and daily activities	Arrange the space adaptable to accommodate activities for all members of multi-family	For economic purpose ignore the opportunity to free from crowding	Use the space sparingly for economic purpose and daily living	Finding the best way to avoid crowding by making the best use of the available dwelling space

CONCLUSION

The idea that crowding has serious consequences for the man appears to have fairly wide acceptance (Altman, 1975; Altman, 1978; Gifford, 2011; Wells, et al, 2007). However, in the case of crowding in the *kampung*, this research suggested that to resist the effect of crowding, *kampung* residents retained immeasurable, intangible variables that should be taken into account. Although crowding involved potential inconveniences, it was not necessarily salient to the perceptions of *kampung* inhabitants. The size of the dwelling space was inevitably inadequate for *kampung* occupants, yet as they had experience living under conditions with limited space, they modified their spatial standards to alleviate the sensation of crowding. Rather than considering crowding as a problem, *kampung* dwellers expressed the benefit of living in dense setting, as it offers each other social support and economic opportunity. This research confirmed Stokols' theory on crowding, that the particular form of one's response to crowding will be a function of the relative intensity of social and personal factors, and the degree to which they can be modified (Stokols, 1972). The findings confirmed that *kampung* inhabitants were comparable with the Chinese and Hispanics who have a greater level of tolerance for overcrowding than Anglo-Americans as a generalization. Many scholars believe that for "close contact" Asian societies, living in confined quarters were judged as being voluntary or at least tolerable (Gove and Hughes, 1983; Hall, 1966; Stokols, 1972).

Crowding in the *kampung* was not merely caused by the scarcity of space, but also inter-generational co-residence under the same roof. To adjust to the limited economic ability of families to set up new households, crowding is synonymous with multi-generational living in the *kampung*. Affordability issues were not the only reason for tolerance levels related to crowding. For the elderly and for those who needs assistance and a quick response from family members (such as in the case of injuries and other accidents), smaller spaces mean more eyes and ears are available. An obligation towards family members, personal preferences, the need for mutual support, or a combination of these factors appeared to be a strong factor in why crowding does not always negatively impact the well being of *kampung* dwellers.

The finding strengthens the need of innovative and alternative method to avoid isolation, simplification and superficial approach in developing the knowledge on housing affordability (Salama, 2007). While in fact the activity of designing contemporary domestic space mostly relies on western spatial assumptions, residents' method of appropriating and thinking about space could not be overlooked. Additionally, the affirmative reaction to the high density of an inhabitant's adaptation level raised the need for further research on residents' preferences in relation to *interiority*. Another path of inquiry is on the wealth of ideas beyond crowding, which here not yet been explored in this research. It is also important to link the findings to other variables of the outdoor space and spatial arrangement of the whole *kampung*, such as the close-knit social network in the *kampung* community and the physical layout that enhancing social cohesion among residents.

ACKNOWLEDGEMENTS

The author thanks to Sherly Listiyanti (junior architect) who helped to organize the survey, supervised surveyors and provided all figures in this paper, and Diane Wildsmith (Visiting Assistant Professor Universitas Indonesia) who assisted in editing the paper.

REFERENCES

Abeyasekere, S. (1987). Jakarta: A History, Oxford: Oxford University Press.
Altman, I. (1975). the Effects of Crowding and Social Behavior, California: Brooks/Cole Publishing Co.
Altman, I. (1978). Crowding: Historical and Contemporary Trends in Crowding Research in Baum, A, and
       Epstein Y.M (Eds.), Human Response to Crowding, Hillsdale, New Jersey: Lawrence Erlbaum
       Associates.

Ambrose, P. (1996b). The Real Cost of Poor Homes: A Critical Review of the Literature, University of Sussex and University of Westminster.

Baker, M., Mc Nicholas, A., Garrett, N., Jones, N., Stewart, J., Koberstein, V., and Lennon, D. (2000). Household crowding a major risk factor for epidemic meningococcal disease in Auckland children, Pediatric Infectious Disease Journal 19(10), 983-990.

Baldassare, M. (1988). Residential crowding in the United States: A review of the research, Handbook of Housing and the Built Environment in the U.S. (eds.) E.Huttman and W. van Vliet.

Bourdeau, P. (Ed.) (1986). the Forms of Capital, New York, Greenwood Press.

Brown, S. C., C. A. Mason, J. L. Lombard, F. Martinez, E. Plater-Zyberk, and A. Spokane (2009). The Relationship of Built Environment to Perceived Social Support and Psychological Distress in Hispanic Elders: The Role of Eyes on the Street. Journal of Gerontology: Social Sciences 64b (2): 234–46.

Certeau, M.D. (1984). The Practice of Everyday Life. Berkeley: University of California Press.

Castels, L. (1967). the Ethnic Profile of Djakarta, Indonesia. 1(April), pp.153-204.

Clark, W.A.V. (1992). Comparing cross-sectional and longitudinal analyses of residential mobility and migration. Environmental Planning A 24 1291-1302

Clauson-Kaas, J., Dzikus, A., Surjadi, C., Jensen, H., Hojlyng, N., Aaby, P., Baare, A. and Stephens, C. (1997). Crowding and Health in Low-Income Settlements, United Nations Centre for Human Settlements (Habitat), England: Avebury.

Colombijn, F. (2010). Under construction: Urban space and housing in Indonesia during the decolonization, 1930-1960. Leiden: KITLV Press.

Crothers, C. Kearns, R. and Lindsey D. (1993). Housing in Manukau City: Overcrowding, Poor Housing and Their Consequences. Working Papers in Sociology No. 27 University of Auckland.

Curry, A., C. Latkin, and M. Davey-Rothwell (2008). "Pathways to Depression: The Impact of Neighborhood Violent Crime on Inner-City Residents in Baltimore, Maryland, USA." Social Science & Medicine 67: 23–30.

Davies. R.B. Pickles, A.R. (1985). Longitudinal versus cross-sectional methods for behavioral research: a first round knockout. Environmental and Planning A17.pp 1315-1329.

Edwards, J., Fuller, T., Sermsi, and Vorakitphokatorn, S. (1994). Household Crowding and Its Consequences, Colorado: Westview Press.

Evans, G. W., N. M. Wells, and A. Moch. (2003). "Housing and Mental Health: A Review of the Evidence and a Methodological and Conceptual Critique." Journal of Social Issues 59: 475–500.

Flick, U. (1998). An Introduction to Qualitative Research. Thousand Oaks, California: Sage.

Freedman, J. L. (1975). Crowding and Behavior. San Francisco: Freeman.

Funo, S. Yamamoto N. and Silas, J. (2002). Typology of *Kampung* Houses and Their Transformation Process, a Study on Urban Tissues of an Indonesian City. Journal of Asian Architecture and Building Engineering/JAABE vol.1 no.2, 193-200.

Funo, Shuji; F Ferianto, Bambang; Yamada, Kyouta (2005). Considerations on typology of *Kampung* Luar Batang (Jakarta). Journal of Asian Architecture and Building Engineering, Vol. 4 (2005) No. 1, 129-136

Funo, Shuji; F Ferianto, Bambang; Yamada, Kyouta (2004). Considerations on typology of *Kampung* Luar Batang (Jakarta). Journal of Asian Architecture and Building Engineering, Vol. 3 (2004) No. 1, 129-136.

Fuller, T.D., Edwards, J.N., Sermsri S. and Vorakitphokatorn, S. (1993). Housing, stress and physical well-being: Evidence from Thailand, Social Science Medicine Vol. 36, 11, 173-180.

Gifford, R. (2011). Environmental psychology. In The IAAP Handbook of Applied Psychology, Edited by Paul R. Martin, Fanny M.Cheung, Michael Knowles, Michael Kyrios, Lyn Littlefield, J. Bruce Overmier, and Jose M. Prieto. London: Blackwell Publishing Ltd.

Gove, W., Hughes, M. and Galle, O (1979). Overcrowding in the home: An empirical investigation of its possible pathological consequences. American Sociological Review Vol. 44 February: 59-80.

Greenfield, R. J. and Lewis, J.F. (1973). An alternative to a density function definition of overcrowding, Housing Urban America J. Pynoos et al (eds.), Chicago: Aldine Publishing Co.

Hall, E. T. (1966). the Hidden Dimension. New York: Doubleday.ks.

Huang, Y. (2003). A room of one's own: housing consumption and residential crowding in transitional urban China. Environment and Planning A 2003, volume 35, pages 591 – 614.

Jazwinski, C. (1998). Crowding http://condor.stcloud.msus.edu/~jaz/psy373/7.crowding.html

Jellinek, L. (1995). Seperti Roda Berputar, Perubahan Sosial Sebuah *Kampung* di Jakarta, Jakarta: PT Pustaka LP3ES Indonesia.

Kato, H. (2012). Report of Preparatory Survey in Cikini. December 26, 2011, Kyoto: Research Institute for Humanity and Nature.

Kent, S. (1990). Domestic Architecture and the Use of Space: an Interdisciplinary Cross-Cultural Study. London: Cambridge University Press.

Krausse, G. H. (1978). Intra-urban variation in *kampung* settlements of Jakarta: A structural analysis, The Journal of Tropical Geography 46: 11-26.

Lawrence, J.E.S. (1974). Science and sentiment: Overview of research on crowding and human behavior, Psychological Bulletin 81,pp. 712-720.

McCarthy, P. (Ed.) (2003). Urban Slums Report: The Case of Jakarta, Indonesia, World-Bank.

Miller, D. (1987). Material Culture and Mass Consumption, Oxford: Blackwell.

Muller, E. J. & Tighe, J. R. (2007). Making the case for affordable housing: Connecting housing with health and education outcomes. Journal of Planning Literature, 21(4), 371-385.

Ranson, R. (1991). Healthy Housing: A practical guide. London: E. & F. N. Spon and WHO.

Rachman, E. (1995). Jakarta, 50 tahun dalam Pengembangan dan Penataan Kota. Jakarta: Pemerintah Daerah khusus Ibukota Jakarta.

Rodgers, W.L. (1981). Density, crowding and satisfaction with the residential environment, Social Indicators Research 10. pp. 75-102.

Ruts, W. (1987). Cities and Towns in Indonesia. Stuttgart: Gebruder Borntraeger.

Salama, A.M. (2015). Editorial: Advandcing the Debate on Architecture, Planning and Built Environment Research, Archnet-IJAR, Volume 9 – Issue 2- July 2015 – (iv-viii).

Salama, A.M. (2007). Navigating Housing Affordability between Trans-Disciplinary and Life Style Theories the Case of Gulf States, Archnet-IJAR, Volume 1 – Issues 2 – July 2007 (57-56)

Sampson, R. (2003). Neighborhood context of well-being. Perspectives in Biology and Medicine, 46 (3), S53-S64.

Schulz, A., William, D., Israel, B., Becker, A., Parker, E., James, S. & Jackson, J. (2000). Unfair treatment, neighborhood effects, and mental health in Detroit metropolitan area. Journal of Health and Social Behavior, 41, 314-332.

Silas, J. (2010). *Kampung* Improvement Program in Surabaya, in D'Auria, V. et.al. (Eds) Human Settlements Formulation and (re)Calibration, Amsterdam: SUN Architecture Publisher.

Silver, C. (2011). Planning the Megacity: Jakarta in the Twentieth Century. New York: Routledge.

Somantri, G.R. (2000). Looking at the Gigantic *Kampung*: Urban Hierarchy and General Trends of Intra-City Migration in Jakarta. Jakarta: FEUI Press.

Sullivan, J. (1992). Local government and community in Java: An urban case study. Singapore, Oxford, New York: Oxford University Press.

Suparlan, P. (1984). Kemiskinan di Perkotaan, Jakarta: Obor.

Stokols, D. (1972). on the distinction between density and crowding: Some implications for future research. Psychological Review, 79(3), 275–278.

Tunas, D. (2008). The Spatial Economy in the Urban Informal Settlement. Netherland: International Forum on Urbanism.

Wells, N. M. & Harris, J.D. (2007). Housing quality, psychological distress, and the mediating role of social withdrawal. Journal of Environmental Psychology, 27, 69-78.

AUTHOR

Evawani Ellisa
Associate Professor, Dr.
Universitas Indonesia,
Department of Architecture Faculty of Engineering
Email address: ellisa@eng.ui.ac.id

7

RESEARCH UTILIZATION IN THE DESIGN DECISION MAKING PROCESS

Amy Huber
Florida State University, Tallahassee, Florida, USA

amattinglyhuber@.fsu.edu

Abstract

This article summarizes findings from a national survey of interior design practitioners in the United States (N=366). The study explored interior design practitioners' current preferences for conducting project research including: activities conducted and sources used, attraction to and recall from sources, and ideas for communicating research findings. Responses suggest that interior designers do value research, yet have little time to utilize research. While cross-tabulation analyses indicate no major differences in research activities between the study's demographic groups, collectively, only 12% of the sample indicated they utilized academic journals and, at times, even incorrectly identified those sources. Open-ended responses allowed designers to offer ideas for communicating research and four key themes emerged, including: topic selection and relevancy, ideas for new dissemination methods, ideas for presentation style, and perceptions of the written language used. It is hoped that this study's findings may help design researchers better communicate their own findings to design practitioners.

Keywords: *Design Research; Research Utilization; Interior Design; Peer Review; Knowledge Acquisition*

INTRODUCTION

A similar scenario unfolds each day: an interior designer is designing a busy social services office. The project's lobby needs to accommodate anxious clients waiting for their appointments. The designer understands this may be a stressful experience. However, he/she is unaware of available research findings on how to mitigate environmental stressors. Even if the designer is aware, he or she feels too busy to read a lengthy journal article, so they briefly discuss options with a colleague, then follow their intuition, and hope for the best outcome. If the best decisions are not made, it will likely be the anxious clients who feel the repercussions.

It has been stated that the format and availability of information are "key variables in the success of design" (Wild, McMahon, Darlington, Liu & Culley, 2010, p. 46). While contemporary discourse has focused on the importance of knowledge, designers with the best of intentions may lack the time and awareness necessary for making informed design decisions. While information is widely available, it is often widespread and decentralized (Mays & Kossayan, 2011). Moreover, interior designers may be working on multiple projects and are concerned with the number of hours billed to their respective clients (Hill, Hegde, & Matthews, 2014). These concerns can truncate research efforts, as emphasis may be placed on the rapid production of contracted deliverables. Given time constraints and multiple project-related stressors, a designer must quickly evaluate an information source in terms of usefulness and decide how much time they will devote to information comprehension.

Previous studies have explored interior designers' perceptions of research (Dickson & White, 1993), its importance relative to professionalism (Birdsong & Lawlor 2001), and its application in design education (Dickinson, Anthony, & Mardsen, 2012). However, studies have yet to explore current methods of designers' knowledge acquisition given the newfound availability of information from internet sources. Further, few studies have been conducted to understand their perceptions of research and their preferences surrounding information sources. This study sought to understand how interior design practitioners utilize research, the sources of information they prefer, and to identify their ideas for communicating research findings. A thorough understanding of these factors might influence the creation of effective formats to disseminate empirical findings to practitioner audiences.

LITERATURE

Perceptions of Research

There are several key studies which have examined interior designers' perception of research, and resulting evidence does suggest practitioners recognize the importance of project-related research. In a quantitative study of interior designers' perceptions of professionalism, Birdsong and Lawlor (2001) found 64.9% of their sample felt research was an *important* component of the design profession.   Dickinson, Anthony, and Marsden's (2012) survey of interior design practitioners indicated that 93% agreed that undergraduate students should know how to use research results during their design process. The same study found younger, more educated interior designers who practiced commercial design were significantly more likely to value research. Yet, in Dickson and White's (1993) survey, only 34% of the sample indicated they researched an interior design problem 100% of the time, and another 27% indicated they researched a design problem less than 25% of the time. These studies suggest the degree to which research manifests in the interior design process likely depends on the scale, scope, and specific needs of the project at hand. While it is unlikely that interior designers perform the types of explorative, open-ended means of *problem seeking* as noted by Maturana (2014, p. 36), the value of research to design is likely increasing in the Knowledge Age. Yet, there remain many variables in the types of knowledge acquisition activities conducted and how information is used.

Research Utilization

The degree to which research influences decisions is commonly referred to as *research utilization* (Wiess, 1979). The literature on the utilization of research spans diverse fields including health, education, and human services (Backer, 1991). Literature suggests three types of research utilization strategies, including: instrumental, conceptual, and symbolic (Pelz, 1978). Instrumental research provides knowledge in specific and direct ways. This type of knowledge would be sought after and applied to a specific problem at-hand. Conceptual strategies involve the use of research for one's general enlightenment.   This type of knowledge is less likely to influence decisions.   Whereas, symbolic utilization strategies involve the use of research to legitimize predetermined notions (Beyer, 1997).

Scholars from the field of environmental and behavioural research have long lamented designers' avoidance of research (Seidel, 1985; Sommer, 1997). To counter this tendency, they offered strategies for increasing research utilization. Seidel (1985) identified three such strategies, including: clarification and dissemination, linkage theories, and collaboration. Clarification and dissemination strategies place emphasis on making relevant information readable and available through presentation style (i.e., graphics, format), approachable language, and the accessible methods for reporting results. Backer (1991 & 1993) suggested that to increase the utilization of knowledge, researchers should seek a user-orientated transformation of knowledge. Such tactics include the elimination of jargon and unnecessary statistics. Going further, linkage strategies suggest the use of an approachable middleman to convey information (Seidel, 1985), and collaboration strategies emphasize the benefits of researchers and research users working together, thus removing communication barriers (Seidel, 1985). While decades have passed since these recommendations were made, it appears little may have changed. More recently, Popov (2009) noted a need for new professional interactions and communication patterns to increase research utilization in design domains. Additionally,

Sommer (1997) suggested factors that may increase research utilization such as prompt and easy implementation of findings, visual appeal, and the use of vivid numbers and verbatim comments. Conversely, factors that may diminish the use of research can include the lack of technical knowledge by viewers (Sommer, 1997). While scholars continue to tout the importance of research utilization, few studies within environmental design research have attempted to identify the characteristics of research knowledge that influence adoption or utilization by design practitioners (Imeokparia, 2005).

Relative to interior design, research suggests varied but often cursory research utilization practices. Some interior designers may strictly utilize existing research, while others may conduct inquiries intended to generate new knowledge. Phares' (2011) survey asked healthcare interior design specialists if they *usually* conducted design research according to predetermined definitions; 79% indicated they followed research literature; 68% measured outcomes; 29% indicated that they shared what they had learned; and only 12% indicated they submitted their findings for peer review. Martin's (2014) interviews of non-healthcare design practitioners suggested that their definitions of research were generally in accordance with information gathering practices. While all her interviewees had some knowledge of Evidence-Based Design (EBD) and all subjects claimed that their firms used research findings, subjects often lacked understanding of actual EBD practices. Dickson and White's (1993) survey indicated that practitioners generally conducted research that reviewed the technical aspects of design (88%), or drew on traditional and past experiences (84%). In fact, only 33% of their sample said they frequently reviewed social or scientific research. This may indicate that while practitioners are apt to use research in their design process, their methods are aimed at instrumental or symbolic utilization strategies. Moreover, they are not likely to conduct original inquiries.

Sources of Information

There are several studies regarding the sources of information interior design practitioners use to acquire knowledge. Findings suggest these sources of information support the pragmatic nature of their research efforts. Dickson and White's (1993) survey indicated information sources most commonly used by designers were product catalogues, design magazines, *Architectural Graphic Standards*, and textbooks. Further, 47% of their sample indicated they *never* consulted scholarly journals for their research. This led the authors to conclude that interior designers have traditionally been reliant on *soft sources* (e.g., periodicals, trade literature) for project-related information (Dickson & White, 1993). This tendency is not unique to interior design. The University of Minnesota's College of Architecture's Dean, Thomas Fisher (2004) jokingly suggested that architects were "allergic to data tables and descriptive statistics" (p. 1), and in her 1999 essay, Susan Roth suggested that much of the knowledge created at universities is not being channelled to industrial designers.

However, several recent cross-disciplinary paradigms have increased the value placed on more rigorous design research efforts. The popularization of the Design Thinking movement (Brown, 2009; Dohr & Portillo, 2011; Dorst 2011; Lockwood, 2009) provided designers with an approachable, yet systematic methodology for collecting data through pre-design observations. Additionally, EBD (Hamilton & Watkins, 2009; Nussbaumer, 2009) strategies have increased the demand for useable empirical evidence as an antecedent to design decision making (Bosch & Nanda, 2011; Cama, 2009). In fact, the aforementioned survey by Phares (2011) found that 72.9% of healthcare designers were "very interested" (p. 51) in evidence-based design practices.

As a result of technology and changing paradigms, there are many information sources available to designers. In addition to scholarly articles, large design firms (Cannon Design, 2013; Gensler, 2014; Perkins & Will, 2013) and contract furniture manufacturers (Herman Miller, 2013; Knoll, 2013; Steelcase, 2013) are engaging internal scholars to generate their own proprietary research. These findings are often disseminated through graphic white papers, presentations, and at trade shows; yet, constraints on time and competitive business conditions may limit submission for external peer review. Professional organizations (IIDA Knowledge Network, AIA Knowledge Net, ASID Knowledge Center) and strategic partnerships (InformeDesign) have attempted to enhance practitioners' connection to academic findings by providing research directories populated with succinct descriptions of empirical studies. However, the extent to which these are currently used by designers is unclear.

While previous studies have highlighted a reliance on soft sources for information, these studies were conducted prior to the wide-spread adoption of the Internet. Thus, it has not yet been determined if the increased availability of research offered by internet sources, or the growing attention to research resulting from EBD and Design Thinking paradigms have altered which sources designers seek in their research efforts.

Scholarly Sources in Design Practice

In order to understand the context of research within the profession of interior design, it is important to recognize the legal requirements for practicing in that field. While these vary by jurisdiction, in the United States, 27 states have requirements that are typically comprised of examination, experience, and training (ASID, 2013), the latter of which typically consists of a 4-year bachelor's degree (NCIDQ, n.d.). This differs from other professions such as medicine and law that do require advanced degrees. Moreover, Bosch and Nanda (2011) noted that only a few design firms have doctoral researchers on staff. As such, an interior designer's exposure to academic literature is likely to be relatively limited as compared to other professions. That being the case, their educational experience may influence their decisions concerning the breadth and depth of their information-gathering efforts.

In his book *Spatial Design Education,* Salama (2015) analysed multiple pedagogical strategies from an array of architecture programs and concluded they generally adopted a research paradigm that placed little importance on developing or analysing current theories. This examination led him to infer design pedagogy as generally emphasizing "high advocacy and low inquiry" (p. 315). Salama's (2008, 2015) design-based 'Trans-Critical' pedagogy theory was offered to address the integration of knowledge across disciplines (Disciplinary Component), the methods by which knowledge is acquired (Inquiry-Epistemic), and how students assimilate new knowledge (Cognitive-Philosophical). Other scholars have proposed tools and methods which emphasize the collection and analysis of information (Bose, Pennypacker & Yahner, 2006; Marinic, 2010; Orthel, 2015; Oxman, 2004), and the integration of knowledge (Deshpande & Khan, 2010). Yet, the pervasiveness of these integrative models in unclear – especially in interior design – and confusion remains amongst students surrounding the role of research in interior design practice (Dickinson, Marsden, & Read, 2007).

In addition to training, other reasons for an interior design practitioner's avoidance of academic sources may be due in part to work pressures and current modes of communication. Journal articles are of course a form of written communication. Yet, in a multiple domain study, Adler, Gujar, Harrison, O'Hara, and Sellen (1998) indicated that linear continuous reading is an "unrealistic characterization" (p. 248) of how people read at work. Their analysis of worker diaries and subsequent interviews found that reading tended to "co-occur" (p. 245) with writing, either in creation of new documents, amending existing documents, or providing clarity to what was being read (e.g., note taking). Their analysis also revealed that "cross referencing" (p. 245) across multiple documents was common. These findings may indicate workers typically apply their reading directly to the task at hand—thus, suggesting instrumental research utilization strategies. While outwardly efficient, this direct application could limit internalization of the information and reduce the likelihood of reading the entire source, tasks which are often necessary in reading academic articles. Further, Wild et al.'s (2010) study of engineering designers' research processes found documents were often utilized as quick references, with 42.1% spending less than 10 minutes examining specific documents such as correspondences, journals, drawings, and standards, among others. Additionally, current journal articles also typically contain relatively little imagery, yet this type of communication is normative to design practice (Ching, 2009; Lawson, 2005). Although methods by which interior designers might leverage visual communication strategies in des gn research has garnered little attention from researchers.

The sources of information used while making design decisions are of utmost importance, and evaluating information sources for validity and appropriateness is imperative (Hamilton & Watkins, 2009). Consequently, it is important to note the nuances between peer-reviewed findings and research studies situated within businesses domains. While many design-related businesses conduct valid and meaningful research, these studies often do not undergo the scrutiny of an external peer review. The peer-review process limits potential bias and efforts are made to share all findings, even those that do not necessarily support the hypothesis of the researcher (Hamilton & Watkins, 2009). Due to competitive factors, this level of transparency would be challenging to replicate within design domains, especially in settings where the research is based on a paid deliverable (i.e., design solution), and sharing any potential project shortcomings can be detrimental to the firm's business development opportunities. Moreover, architecture and design firms are not afforded the types of protections and incentives offered by U.S. patent law (Levin et al., 1987), nor do they have access to the non-biased funding agencies that are available to health and science disciplines. Collectively, these factors potentially deter their peer review efforts.

As such, there are likely advantages for practitioners who use peer-reviewed research. First and foremost, they would be equipped with non-biased foreknowledge to inform their design decisions. Secondly, their more informed design decisions may lead to improved outcomes which may have long-lasting, positive influences on their clients and end users. Third, the profession of interior design itself would likely be advanced by application of research findings.

Summary

While literature suggests that interior designers do value research, they may hold inaccurate perceptions of what research entails, and subsequently be unlikely to utilize empirical research findings. Moreover, literature suggests that the information sources used by interior designers are often pragmatic in nature and could be considered indicative of instrumental and symbolic research utilization strategies. While strategies to increase research utilization have been offered, including clarifying findings and offering new dissemination channels, these recommendations have yet to be implemented and tested in design disciplines. Thus, effective clarification and dissemination strategies for those conducting instrumental research has yet to be documented; in their absence, persuading designers to enact conceptual research strategies is likely difficult. While a few studies have been conducted to better understand interior design practitioner perceptions of research and sources used, these are becoming dated. Moreover, researchers have yet to determine how these designers utilize and conduct research, and their preferences for receiving information, especially given the enhanced availability of information sources. Thus, further study is needed.

METHODS

This study sought to understand interior design practitioners' current preferences for conducting and utilizing research by establishing baseline data regarding the way in which designers currently conduct design research-orientated tasks. It was grounded in the following research questions:

- What types of research are conducted?
- What sources of information do practitioners utilize?

- What preferences do they have regarding information sources (attraction and recall)?

The survey utilized an online questionnaire (Qualtrics) for data collection, took less than 20 minutes to complete, and consisted of two sections. The researcher obtained approval for the study, HSC # 2014.12212, by the Institutional Review Board at Florida State University on March 13, 2014.

Survey Design

As this phase of data collection was exploratory in nature, no previously generated survey instruments were deemed appropriate, thus necessitating a new instrument. To minimize possibility for error, this instrument was reviewed by two distinguished researchers, a statistical consultant, American Society of Interior Designers' Director of Market Research, and subsequently pre-tested by three separate groups of either researchers or design practitioners.

The first portion of the survey contained demographic questions. The second section of the survey began by asking respondents if they conducted project related research. The following definition was used for research "the identification of important design questions and the development and use of organized problem-solving methods" (Thompson, 1992 p. 47). While more contemporary definitions have been offered, this was used by the researcher as it included more pragmatic approaches and did not focus on the generation of knowledge. This was important as definitions and perceptions of design research vary (Dickinson et al., 2007) and previous studies have indicated designers do not typically associate research with original discovery (Dickinson et al., 2012). If participants indicated they did not conduct project related research they were asked reasons why and exited the survey. Those that indicated they conducted research answered questions about their research activities and continued to the final section of the survey which queried how they processed information. Survey items were both quantitative and qualitative and several items allowed for open-ended responses to expand upon answers.

Validity and Reliability

Statistical checks of reliability included inter-rater reliability for open-ended responses and a check for overall response agreement between practitioner pre-test and final survey responses. The researcher sought to establish content, predictive, and construct validity (Creswell, 2009, p. 149) through the writing of the questions themselves and subsequent series of pre-tests and revisions. Whenever appropriate, questions allowed for both closed and open-ended responses to test for predictive validity. However, to maintain overall survey brevity and attain a high response rate, there was limited retesting of items. The pre-test responses by the practitioner group allowed for establishing predictive validity by examining responses against previously published research findings where possible, and comparing their responses to known information about their research practices. Following final revisions to the instrument, construct validity was verified through a final crosscheck of the instrument with the research questions.

Sampling

The target population for the study was interior designers who are actively involved in design projects within the United States. The participants were recruited from membership lists of the American Society for Interior Designers. This organization was selected because it is the oldest professional organization for interior designers in the United States, has the largest body of membership, and it traditionally represents designers practicing both residential and commercial design. A recruitment email was sent to a random sample of 6849 Allied[1], Associate[2], and Professional[3] members. These membership types infer that the designer has met necessary requirements to qualify for these levels. Thus, using these membership types helped the research to better target practicing designers—and filter out responses from interior design educators, product representatives, and students. The invitation email included a link to the survey. After the initial email was sent to the designated sample, two email reminders were sent to addresses of those who had not yet responded. The only identifier to each completed questonnaire was an IP address, unless participants offered to share their email addresses for a follow-up interview.

Survey Analysis

Responses to closed-ended questions were analysed using descriptive statistics (e.g., frequencies and percentages). Inferential statistics were used to understand if demographic characteristics were associated to varying types of research activities conducted and research preferences. This examination included: cross tabulation analysis with *Chi*-square statistics, and ANOVA tests. Additionally, when ANOVA distribution and variance assumptions were confirmed, a post-hoc Tukey's Range Test was used to determine specific differences for those variables where significant p-values were calculated. Open-ended responses were inductively coded by keyword and grouped by theme, a second reviewer then coded the responses and inter-rater reliability was deemed sufficient (<.7) using Cohen's Kappa.

FINDINGS

Three hundred and sixty-six ASID members responded to the survey (a response rate of approximately 5.3%). Fifty-nine percent of participants were aged 51 or above. Sixty-three percent of respondents had over 10 years of professional design experience. Forty percent reported that they were principals/owners of their firm, and 47% reported they worked in a sole proprietorship. Participants were primarily residential designers (67%), and most respondents held a design-related Bachelor's degree (64%), while 18% held an advanced degree. Table 1 provides a summary of demographic information from the survey respondents. While the respondents represented a reasonably large sample population for interior design, it is important to note that they represented a relatively large percentage of older, residential practitioners, who owned their own firms; hence, generalizing this data to the entire spectrum of interior design professionals is inappropriate.

---

[1] ASID Allied Membership: "practicing interior designers who have completed 40 semester or 60 quarter credit hours in interior design education from an accredited institution" (ASID, n.d.a. para 1)

[2] ASID Associate Membership: "practicing interior designers who can demonstrate six years of full time interior design experience and provide a college transcript reflecting a minimum of an associate's degree in a subject other than interior design" (ASID, n.d.b, para 5).

[3] ASID Professional Membership: "requires proof of passage of the NCIDQ examination. (ASID, n.d.c, para 1).

Table 1: Sample demographics

Variable	Response	%
Current age (n=358)		
20-30 years	57	16
31-50 years	93	26
51-65 years	163	46
Over 65 years	45	13
Years of experience (n=358)		
Less than 2	35	10
2-5	43	12
6-10	54	15
11-20	68	19
Over 20	158	44
Current position (n=358)		
Junior Designer/Architect	26	7
Interior Designer/Architect	105	29
Senior Designer/Architect	38	11
Design Director	7	2
Project Manager	7	2
Principal/Owner	143	40
Other	32	9

Other positions included: Project coordinators, program managers, product development, librarians, consultants, sales, and managers

Variable	Response	%
Size of firm (n=350)		
Sole Proprietorship	164	47
2-5 Designers	124	35
6-20 Designers	40	11
21-50 Designers	11	3
51-200 Designers	6	2
Over 200 Designers	5	1
Primary Market Sector (n=356)		
Commercial/Corporate Interiors	58	16
Health & Wellness	15	4
Residential	237	67
Retail/Hospitality	15	4
Education	10	3
Other	21	6

Other included: Food & Beverage, Senior Housing, Textile Design, Government, Historic Preservation, Aviation, Collegiate Bookstores

Variable	Response	%
Level of Education (n=352)		
Certificate	25	7
Associates Degree	36	10
Bachelor's Degree	225	64
Master's Degree	61	17
Ph.D.	5	1

Non-design related Bachelor's degrees included: *Marketing, Art, Journalism, Business, Visual Communication, Textile Marketing, English, Accounting, Business Management, Early Child Education, Communication, Political Science, Fine Art, Advertising, Public Relations, Psychology, Religious Studies, Industrial Design, Graphic Design, Business Education, Sustainable Management, English*

Non-design related Master's degrees included: *Project Management, Speech Pathology, Business Administration, Educational Psychology, Gerontology, International Art & Architecture, Early Childhood Education, Social Work*

Non-design related PhD. Degrees: *Allied health*

Research Activities

Following demographic questions, participants were asked questions regarding their current research practices. Eighty-nine percent of respondents indicated they conducted design research in accordance with the Asher Thompson definition.

Table 2 illustrates the types of research tasks the respondents conducted, when multiple answers are permitted. *Analysis of design trends, product research or prototyping, and client-based research,* received the greatest number of responses. Generally, these responses suggest that most of the research tasks conducted are application-based, focusing on trends, client-based information gathering, and product specifications. Conversely, on-site observations and studies of human behaviour were indicated by relatively few participants. This may suggest that the research types conducted represent areas where the designer may already have high familiarity and are normative to their regular activities. In the open-ended prompts, respondents also indicated conducting research focusing on aging in place solutions, and trends focused more specifically on product development.

Table 2: Types of research conducted

Type of research conducted (multiple responses allowed)	n=307	%
*Analysis of design trends *Monitoring works of competing firms*	228	74

*Product research or prototyping
    Reading product literature
    Building Fabricating Mock-ups
	211	69

*Client-based research *Examining client's business plan, goals, or lifestyles to inform programming*	203	66
Sustainability *Exploring better systems, methods or product specifications with a goal of sustainable outcomes*	166	54
*Monitoring business & lifestyle trends *Exploring broad topics which may influence design decisions*	166	54
Interviews and/or Focus Groups with project stakeholders *Meeting with key stakeholders to better understand project requirements*	147	48
On-site Observations *Observing patterns, preferences, workflows*	106	34.5
Studies of human behaviour & Environmental Psychology *Exploring relevant psychological, behavioural, or sociological phenomena*	105	34
Post Occupancy Evaluations *Formalized research regarding project successes & failures after project completion*	96	31
Other *Aging in place, product development*	4	1

Comparison of Research Types

As indicated on Table 3, *Chi*-square analysis of demographic information against key variables including: type of research conducted, sources of information, attributes which draw attention, and time allocation indicated relationships only between age and attributes which draw attention (i.e., source of attraction), and between market sectors to type of research conducted.

Table 3: Demographic associations

Variable	Chi-Square	p-Value
Age		
Type of research conducted	16.88	.97
Sources of information	19.62	.55
Attraction to sources	23.94	*.02
Time allocation	17.15	.51
Years of Experience		
Type of research conducted	29.89	.88
Sources of information	19.05	.90
Attraction to sources	13.34	.65
Time allocation	28.81	.23
Size of Firm		
Type of research conducted	38.99	.87
Sources of information	21.79	.96
Attraction to sources	12.28	.91
Time allocation	34.00	.28
Market Sector		
Type of research conducted	81.32	*<.001
Sources of information	23.17	.94
Attraction to sources	9.56	.98
Time allocation	26.34	.66
*p-values significant at the 0.05 level		

ANOVA analysis indicated a significant difference in the mean number of research types across all market sectors ($p=<.001$). A post hoc Tukey's range test was then used to test for multiple comparisons. As shown on Table 4, on average commercial designers conducted more types of research than residential designers ($p=<.001$), and those practicing in education ($p=.17$). Further, retail designers conducted more research activity types than residential designers ($p=.035$), and somewhat more than educational designers ($p=.41$).

Table 4: Types of research conducted *(Measured in Quantity of types of research performed)*

Market Sector	N	Mean	Std. Dev	Std. Error	95% Confidence Interval for Mean	
					Lower Bound	Upper Bound
Commercial	58	5.21	3.013	.396	4.41	6.00
Retail/hospitality	15	5.47	3.091	.798	3.76	7.18
Residential	237	3.30	2.554	.166	2.97	3.63
Health & Wellness	15	4.53	3.182	.822	2.77	6.30
Education	10	2.20	2.394	.757	.49	3.91
Other	21	4.00	3.271	.714	2.51	5.491

Sources of Information

Designers were asked to specify the types of sources used during their research, allowing for multiple answers. As shown in Table 5, respondents typically utilized non-scholarly information sources; only 12% indicated they used Academic Journals. Interestingly, when those who indicated they did read academic journals were asked to share titles of those used, respondents often listed non-peer reviewed sources such as: *New York Times, Interior Design Magazine*, and textbooks. This suggests interior designers may inaccurately classify publications as having gone through peer-review. Respondents indicating *other* sources of information listed several internet based sources (e.g. blogs, and daily email blasts) and catalogues, in addition to product specifications and internally generated sources produced by their firms.

Table 5: Sources of information

Sources of information (multiple responses allowed)	n=305	%
Product literature	270	88.5
Attend conferences/tradeshows/CEUs	255	84
Professional organizations, knowledge-focused web sites	226	74
Discussing topics with colleagues	205	67
Trade publications	192	63
*Academic journals	38	12
Other	34	11
General Internet Searches		
Firm's own research/library		
Online Magazine, Catalogues, or forums		
Company representatives		
*indicates scholarly source		

Respondents who did not indicate using academic journals for research were asked reasons for their avoidance. Many respondents indicated lack of knowledge or lack access to relevant academic journals. Time constraints were also attributed to their avoidance. Yet, 9% of respondents indicated *other* reasons including: not knowing which journals would be relevant, an indication that journal articles were too long, a preference for other sources, or a general perception that academic research topics were irrelevant and either: too vague, limited, or "overly academic" in nature. One respondent indicated: "My perception of them [academic journals] is that the information would not be a quick real world solution and therefore a waste of my time." Few respondents indicated they did not understand how to process the information from academic sources.

Table 6: Reasons for not using academic journals

Reasons (multiple responses allowed)	n=260	%
Does not know about relevant journals	77	30
Time constraints	63	24
Unsure how to find or access relevant journals	46	18
Does not have access to relevant journals	46	18
Other	24	9
Not sure which are relevant		
Competition from other sources		
Forgetting them		
Do not need them		
Topics		
Article length		
Unsure if they use them or not		
I do not understand them	4	1.5

Preferences for Information Sources

Interior designers were asked about their preferences for information sources in terms of their attraction to sources and what they could later recall from information sources. This was used to help determine how interior designers judge an information source and determine its worth.

Attraction

Survey participants were asked what specific attributes that would attract them to sources of information (n=238) responded, allowing for multiple answers. Responses outlined in Table 7 indicate that after topic, graphics, and the source of the article were deemed important.

Table 7: Attraction to specific information sources and processing tools

Attributes that attract design practitioner attention (multiple responses allowed)	n=238	%
Topic of the article	222	93
Graphics contained in the article	119	50
Source in which the article was published	100	42
Pre-existing knowledge of the authors	35	15
Other reasons (themes listed below)	8	3
Use of humour in writing		
Catch phrases & captions		

Recall

In an effort to understand the types of information most likely to be recalled, survey respondents were asked what kinds of information they are most likely to be remember the next day on a scale ranging from 1=Strongly disagree (won't remember) to 4=Strongly agree (will remember) (see Figure 1). Respondents felt most likely to remember big ideas and conclusions, research stories or methods, and specific images and graphics. Conversely, respondents indicated they were most likely to *not remember* details such as statistics.

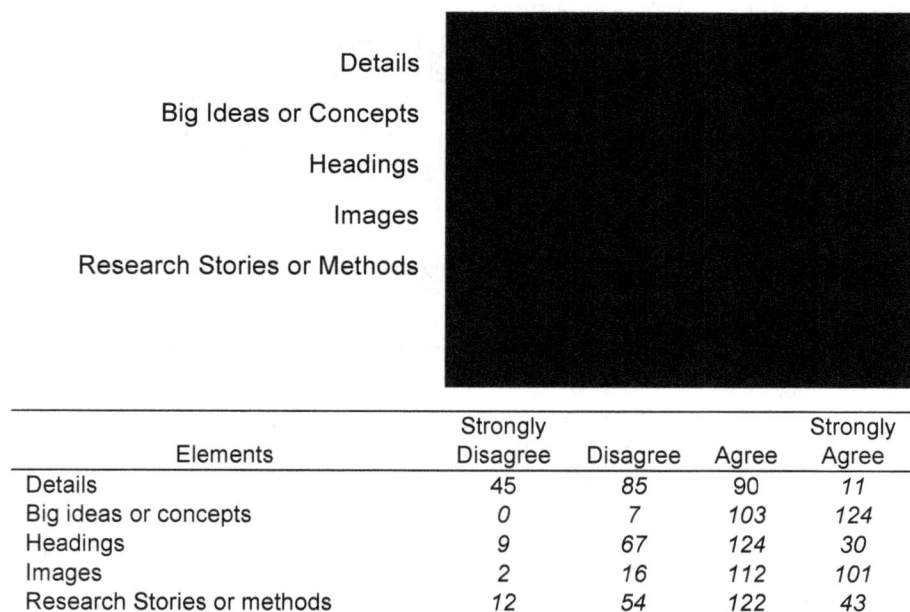

Elements	Strongly Disagree	Disagree	Agree	Strongly Agree
Details	45	85	90	11
Big ideas or concepts	0	7	103	124
Headings	9	67	124	30
Images	2	16	112	101
Research Stories or methods	12	54	122	43

Figure 1. Types of information most likely to be recalled.

Time Allocation to Specific Sources

Designers were asked how much time they would devote to reviewing specific sources of information. As indicated on Table 8, of 279 respondents, 49% percent indicated a time span of less than 10 minutes. Those who responded that their time allocation may vary, indicated they would decide how much time to spend based upon perceived topic relevance, article length, and written style.

Table 8: Time Allocation to Sources

Time spent	n=279	%
Less than 10 minutes	136	49
10-20 minutes	50	18
More than 20 minutes	19	7
Depends on the source	33	12

Qualitative Responses

The practitioners were also asked their ideas for how research is used or may be received. Respondents answered this question in one of two ways; either how they would best receive information or how they would use information during their design process. As indicated on Table 9, emerging themes included: topic selection, dissemination methods, presentation style, and written language. The relatively high quantity of responses in Themes 1 & 2 (i.e., topic selection and dissemination methods) may indicate that barriers in accessing research and that usability of the research are common concerns amongst interior designers.

While elaborating on their responses to dissemination ideas, some designers further demonstrated issues with accessbility of academic research. In fact, some stated using proprietary sources in the belief that these were peer-reviewed. For example: "In my experience, the most accessible academic research on corporate design can be found on major furniture manufacturer's websites such as Steelcase, Knoll, and Herman Miller." Others commented more directly on the presentation style, for example:

> Many designers are visual people, white papers with lots of text and few images will discourage engagement of information. Personally, I prefer text that's supported with images/graphics. I also find it much easier to get information from a film than a book.

Still others briefly mentioned how they judged sources of information, inferring that the social context may influence research habits, "Designers are networkers, and if one designer recommends a specific research tool, that is typically the one I will use". Finally, other designers indicated problems in having to pay for information access.

Last, designers were simply asked if there was anything else they would like to share regarding research style and preferences. While the broad nature of this question does make inductive coding difficult, many of the responses were illuminating in terms of the perceived deep and ongoing disconnect between academic research and practice-based needs.

> For example:
> There is a lot of research that seems pointless to the ordinary person. Unfortunately, this kind of work undermines the perception that research adds something useful to design knowledge.

and,

I realize that every profession has its jargon and academic writing can be filled with important data that is mainly for other academicians. However, if I, as a practitioner, can't glean what I need in a clear and coherent way, and in an efficient way, I am not going to waste my precious time on it.

Others used the opportunity to provide communication ideas,

As a designer, most everything is visual. So, an image, diagram, or drawing of something is far more informative. Sometimes the author is not clear, or the wording is complex. Keep it simple.

Table 9: Themes from Ideas for Communicating Research (n=71) K= .029

Theme 1-Topic Selection & Use of Research (average *f*=29)
Subthemes-listed in order of highest frequency
Resolve issues with usability of research
Resolve issues with purpose of research
Provide tangible examples
Seek interdisciplinary scope
Designers may use of research to (see below):
Improve client relations
Gain ideas
Enhance projects
Confirm priori ideas
Theme 2-Dissemination Methods (average *f*=24.5)
Subthemes-listed in order of highest frequency
Remove barriers to access and increase availability of research
Provide interactive options
Utilize professional organizations
Utilize trade journals
Create CEU classes
Utilize video
Create discussion groups /networks
Create a physical design library
Theme 3-Presentation Style (average *f*=8.5)
Subthemes-listed in order of highest frequency
Increase visual components
Use of keywords
"Presentability"
Theme 4-Written Language (average *f*=6.5)
Subthemes-listed in order of highest frequency
Use concise writing
Quality perceptions
Elucidate clear results & outcomes
Use of bullet points
Uncategorized (*f*=3)

LIMITATIONS

As with any exploratory research design, these findings have limitations. These include: lack of a previously tested survey instrument, and limited internal consistency testing. Additionally, the respondents represented older, residential designers who owned their own firm, thus limiting generalizability to all interior design populations.

As with any survey, situational influences are lacking and respondents may not have answered questions according to their actual task performance. Future research should seek other data collection methods to ascertain how practitioners conduct design research – including on-site observations and protocol studies – as they may allow for contextual nuances. Also, a content analysis comparing attributes of "soft sources" against academic journals may be useful in clarifying any distinctions which may deter design practitioners from using academic sources. Last, experimental research should seek to understand what, if any, physical attributes (e.g., colour, layout) might best garner the attention of design practitioners while conveying the appropriate meaning of information (i.e., message comprehension).

DISCUSSION

Despite the fact that, 23 years ago, Dickson and White (1993) observed that "interior design educators and practitioners must change their perceptions to achieve greater understanding and respect for research and the generation of new knowledge" (p. 10), findings from this study may suggest little has changed. However, while this study confirms earlier literature, it may also provide insight into a means to achieve this change by establishing interior designers' preferences for acquiring knowledge by ascertaining the types of information they currently access, what attracts them to an information source, what they then recall about the source and, lastly, their ideas regarding design research. Collectively, this information may enhance research utilization within design practice.

Perceptions of Research

While findings suggest the age of participants may influence what attracts them to specific information sources and that some market sectors conduct more types of research activities, it should be noted that, overall, these interior designers were inclined to value design research efforts. Eighty-nine percent of respondents indicated they conducted some type of research-orientated activity. This suggests that interior design practitioners do value these activities and supports earlier studies by Birdsong & Lawlor (2001), Dickinson, Anthony, and Marsden (2012), and Martin (2004). However, some of those surveyed surmised that the majority of academic research was not useful to interior design practice. This may imply that research translation efforts continue to be lacking and supports Sommer's (1997) assertions surrounding the importance of prompt and easy implementation of findings.

RESEARCH UTILIZATION

Respondents indicated they were most likely to conduct tasks involving *analysing design trends*, *client-based research*, and *product research*. From these types of research-orientated activities, it could be inferred that designers would be apt to conduct instrumental research utilization strategies (i.e., those applied directly to the task at hand) or symbolic research utilization strategies (i.e., those supporting their own pre-existing paradigms). Moreover, open-ended responses supported Seidel's (1985) recommendations regarding the clarification of research and the necessity for designer-orientated dissemination strategies. In this study, participants noted specific clarification strategies such as enhancements to visual presentation and written style, while their suggestions for enhancing dissemination included removing barriers to access, providing interactive research documents, and finding additional channels with which to share information (e.g., trade periodicals, CEU, and conferences).

Participant responses also suggested that the range of research and information gathering activities in which they most frequently engaged was largely pragmatic in nature, with a focus on the application of knowledge and not its creation. These findings support earlier studies (Dickson & White, 1993) and could likely be attributed to a practitioner's current work environment and their possible need to directly apply information to a project at hand. Further, design practitioners may not recognize a benefit from theoretical research projects that do not yield explicit implications and practical recommendations.

Respondents indicated a preference for graphics in terms of both attraction to and recall of respective information sources. Additionally, when solicited via open-ended questions, their ideas could lead one to conclude that more attention needs to be given to both the format of research documents and how they are disseminated. Since the designers surveyed did note an interest in conducting research and information gathering activities, one could also infer that they may find benefit from newfound sources, including those that have been peer reviewed, given the applicability of information and their ease in obtaining it.

Sources of Information
Dickson and White's (1993) findings related to practitioner reliance on soft sources were supported by this study. Responses suggested that this is due to several factors: time pressures, a perception of topic irrelevancy, presentation style, and language used. Additionally, 30% of this survey's sample indicated they were unaware of relevant journals and 18% did not know how to find them. This may suggest that when available, practitioners would prefer to use peer-reviewed information; however they may be willing to settle for what's easily available, or what they may have time to obtain and review.

When comparing survey responses to studies conducted in allied disciplines, several commonalities emerge. The confounding responses received when asking which academic journals are used may indicate that Roth's (1999) statement identifying industrial designers' uncommon use of scholarly sources may also be true for interior designers. Additionally, survey responses relative to time allocation on specific sources align with some findings from Wild et al. (2010). In their sample, 42% of engineers spent less than 10 minutes with a specific document, and 22% indicated 10-20 minutes, while 38% indicated spending over 60 hours. However, in their study, their range of documents included design drawings, which was not a focus of this study. Given the general length of academic journal articles, these responses may indicate that designers generally lack the time required to process information from such articles without aid from a condensed summation.

IMPLICATIONS
In his aforementioned book, Salama (2015) posited the design process as a merging of "intuition, experience, and the application of skills and knowledge gleaned from traditional training methods" (p. 115). As such the design process is shaped by many factors and that, collectively, these factors influence how designers utilize research when making design decisions. In this case, while the perception of research was seemingly important to the respondents, given the fact that few referred to academic literature or knew how to find these sources, suggests the validity of their sources was seemingly of little concern. This could be due to their educational background or the communication strategies with which empirical studies are traditionally disseminated. However, these findings may provide direction and suggest an effective two-part strategy, one focused on integrative educational paradigms and a second on generating practitioner-orientated research documents.

EDUCATIONAL STRATEGIES

In Dickson and White's (1993) aforementioned article, they stated that the primary role of interior design educators was to "advance the profession through the generation of research that adds to the body of knowledge, to place this research into a contextual framework that can be used by the design profession, and to convey the existing body of knowledge to students" (p. 10). Yet, it appears that educators may still not be placing information into a usable contextual framework. As such, pedagogical models that focus on the integration of knowledge should be emphasized, as should helping students identify, analyze, and utilize peer-reviewed research findings.

On the whole, survey results indicated practitioners are not educated as to the role of scholarly research and the peer review process. While interior design educators do advocate the use of research in studio projects (Dickinson et al., 2009), students can still be reliant on soft sources of information (Dickinson et al., 2007). If these students are graduating from school thinking that browsing the Internet is research, it is then understandable that as practitioners, they would still not know what academic journals are.

The author of this study would like to offer the following recommendations for design educators:

- Seek pedagogical models that focus on the holistic integration of knowledge.
- Provide more experiences that help students identify academic journals; recognize their value, and how to locate these sources.
- Seek ways to expose students to post occupancy evaluation methods and the benefits of sharing these findings beyond the project itself.
- Refrain from inappropriately using the term *research*.
- Discuss issues surrounding bias and the importance of validating sources.

Yet, there are many interior designers who were taught under a paradigm less focused on research, and educating them on the benefits of research is also important, as is communicating research findings in a manner that will be perceived as approachable and accessible.

PRACTITIONER-ORIENTATED RESEARCH DOCUMENTS

Findings from this study suggest the need for a continued and ongoing dialogue surrounding best practices for translating research findings. Researchers interested in sharing findings with design practitioners may look to create two sets of documents: one academically oriented, with the goal of communicating replicable research methods, and one practice-oriented, aimed at sharing applicable findings in a manner allowing for rapid evaluation and application. This author is not advocating for a decrease in the level of research rigor, but rather to effectively share applicable findings with those who may be able to use them directly within the design of built environments.

In this case, recommendations for design scholars would include:

- Create audience-focused research products.
  - *Consider one academically-oriented and one practice-oriented.*
- Produce documents that balance imagery with words
  - *This may allow for more effective information processing by those with little time to process information or for those who frequently communicate via drawings and sketches.*
- Find ways to layer information.
  - *Explore media types that may allow choice in the intensity of information engagement.*
- Illustrate potential connections between empirical findings and project application.
  - *This may include more use of contextual graphics (diagrams, project imagery) and succinct, pragmatic design suggestions.*

- Consider how designers may communicate research findings to their own clients.
  - *Designers may be more apt to use information that can be easily shared.*
- Consider the approachability and availability of findings.
  - *Acknowledge that there is a multitude of information options available to practitioners, and knowledge of how one comes to find information and validate sources is important.*

SUMMARY

While this study is exploratory and only a first step in a multi-faceted research plan, its intent is to further discussions aimed at lessening the perception of a deep and ongoing rift between practice-based design processes and peer-reviewed research findings. This is an important step, as the goal of design research should be to improve the design of space, thus encouraging responsive environments and enhancing the lives of those using these spaces.

REFERENCES

Adler, A., Gujar, A., Harrison, B., O'Hara, K., & Sellen, A. J. (1998). A diary study of work-related reading: Design implications for digital reading devices. *Proceedings of CHI '98 (Conference on Human Factors in Computing Systems),* Los Angeles, CA. Retrieved from http://www.interactiondesign.org/references/conferences/proceedings_of_the_acm_chi_98_human_f actors_in_computing_systems_conference.html

ASID. (2013). ASID Legislative Policy. Retrieved from https://www.asid.org/custom/ASID2013/documents/RegistrationLawsMap_1207.pdf

ASID. (n.d.a) Allied Membership qualifications. Retrieved on January 19, 2016 from https://www.asid.org/content/allied-membership#.Vp6_-cL2aUk

ASID. (n.d.b) Associate Membership qualifications. Retrieved on January 19, 2016 from https://www.asid.org/content/associate-membership#.Vp7AdsL2aUI

ASID. (n.d.c) Professional Membership qualifications. Retrieved on January, 19, 2016 from https://www.asid.org/content/professional-membership#.Vp7AjML2aUk

Backer, T. E. (1993). Information alchemy: Transforming information through knowledge utilization. *Journal of the American Society for Information Science, 44*(4), 217-221. doi: 10.1002/(SICI)1097-4571(199305)44:4<217::AID-ASI9>3.0.CO;2-D

Backer, T. E. (1991). Knowledge Utilization: The Third Wave, *Knowledge: Creation, Diffusion, Utilization, 12*(3), 225-240.

Bosch, S., & Nanda, U. (2011). Outside the ivory tower: The role of healthcare design researchers in practice. *Journal of Interior Design, 36*(2), v–xii. doi: 10.1111/j.1939-1668.2010.01055.x

Birdsong, C., & Lawlor, P. J. (2001). Perceptions of professionalism: Interior design practitioners working for the top 100 firms. *Journal of Interior Design, 27*(1), 20-34. doi: 10.1111/j.1939-1668.2001.tb00363.x

Beyer, J. M. (1997). Research utilization: Bridging a cultural gap between communities. *Journal of Management Inquiry, 6*(1), 17-22. doi: 10.1177/105649269761004

Bose, M., Pennypacker, E., & Yahner, T. (2006). *Enhancing critical thinking through "independent design decision making" in the studio. Open House International, 31(3), 33-42.*

Brown, T. (2009). *Change by design.* New York, NY: Harper Collins.

Cama, R. (2009). *Evidence-based healthcare design.* Hoboken, NJ: John Wiley & Sons, Inc.

Cannon Design. (2013, September). *Research.* Retrieved from http://www.cannondesign.com/news/research/

CIDA. (2014). *Future vision report.* Retrieved from http://accredit-id.org/wp content/uploads/2014/10/FV_012315.pdf.

Ching, F. (2009). *Architectural graphics.* (5th ed.). New York, NY: Wiley.

Creswell, J. W. (2009). Research design: Qualitative, quantitative, and mixed methods approaches. Los Angeles, CA: Sage.

Deshpande, S. A., & Khan, A. R. (2010). Towards total integration in design studio. *Archnet-IJAR: International Journal of Architectural Research, 4*(2/3), 61-75. Retrieved from http://www.archnet-ijar.net/index.php/IJAR/article/view/109/117

Dickinson, J. I., Anthony, L., & Marsden, J. P. (2012). A survey on practitioner attitudes toward research in interior design education. *Journal of Interior Design, 37*(3), 1-22. doi: 10.1111/j.1939-1668.2012.01078.x

Dickinson, J. I., Anthony, L. and Marsden, J. P. (2009), Faculty Perceptions Regarding Research: Are We on the Right Track?. *Journal of Interior Design, 35:* 1–14. doi: 10.1111/j.1939-1668.2009.01024.x

Dickinson, J. I., Marsden, J. P., & Read, M. (2007). Empirical design research: Student definitions, perceptions, and values. *Journal of Interior Design, 32(2),* 1-12. doi: 10.1111/j.1939-1668.2006.tb00309.x

Dickson, A. W., & White, A. C. (1993). Are we speaking the same language? Practitioners' perceptions of research and the state of the profession. *Journal of Interior Design, 19*(1), 3-10. doi: 10.1111/j.1939-1668.1993.tb00146.x

Dohr, J., & Portillo, M. (2011). Design thinking for interiors: Inquiry, experience, impact. New York, NY: Wiley.

Dorst, C. H. (2011). The core of 'design thinking' and its application. *Design Studies, 32*(6), 521-532. doi: 10.1016/j.destud.2011.07.006

Fisher, T. (2004). Architects behaving badly: Ignoring environmental behavior research. *Harvard Design Magazine, 21,* 1-3.

Gensler. (2014). *Research catalogue.* (Vol. 1), San Francisco: Gensler.

Hamilton, D. K., & Watkins, D. H. (2009). *Evidence-based design for multiple building types.* Hoboken, NJ: John Wiley & Sons, Inc.

Herman Miller. (2013, December). *Research Topics.* Retrieved from http://www.hermanmiller.com/research/topics/all-topics.html

Hill, C., Hegde, A. L., & Matthews, C. (2014). Throwing in the towel: Burnout among practicing interior designers. *Journal of Interior Design, 39,* 41-60. doi: 10.1111/joid.12029

Imeokparia, T. (2005). The design, implementation, and evaluation of an interactive multimedia environmental design research information system: Architectural design review as case study. (Unpublished doctoral dissertation). Ohio State University, Ohio.

Knoll. (2013, December). *Workplace research library.* Retrieved from http://www.knoll.com/research

Lawson, B. (2005). *How designers think: The design process demystified* (4th ed.). Burlington, MA: Elsevier.

Lawson, B. R., & Dorst, C. H. (2009). *Design expertise.* Oxford: Architectural Press.

Levin, R. C., Klevorick, A. K., Nelson, R. R., Winter, S. G., Gilbert, R., & Griliches, Z. (1987). Appropriating the returns from industrial research and development. *Brookings Papers on Economic Activity, 1987*(3), 783-831. doi: http://doi.org/10.2307/2534454

Lockwood, T. (2009). *Design thinking.* New York, NY: Allworth Press.

Martin, C. S. (2014). Implementation of Evidence-Based Design (EBD) by non-healthcare design practitioners. *International Journal of Architectural Research, 8*(3), 166-180.

Marinic, G. (2010). Bubble up: Alternative approaches to research in the academic architecture studio. *Archnet-IJAR: International Journal of Architectural Research 4*(2/3), 61-75. Retrieved from archnet-ijar.net/index.php/IJAR/article/download/97/104

Maturana, B. (2014). Where is the 'problem' in design studio: Purpose and significance of the design tasks. *International Journal of Architectural Research, 8*(3), 32-44. Retrieved from www.archnet-ijar.net/index.php/IJAR/article/view/466/pdf_

Mays, E. M., & Kossayan, C. (2011). Design research in a non-linear world: Adapting to a ubiquitous playground of content. *Proceedings of the Design Principles and Practices Conference,* Rome, Italy. Retrieved from http://2011.designprinciplesandpractices.com/sessions/index.html

NCIDQ. (n.d.). Eligibility requirements. Retrieved from http://www.ncidqexam.org/

Nussbaumer, L. (2009). *Evidence-based design for interior designers.* New York, NY: Fairchild.

Orthel, B. D. (2015). Implications of design thinking for teaching, learning, and inquiry. *Journal of Interior Design, 40,* 1-20. doi: 10.1111/joid.12046

Oxman, R. (2004). Think-maps: Teaching design thinking in design education. *Design Studies, 25,* 63-91. doi: 10.1016/S0142-694X(03)00033-4

Pelz, D. C. (1978). Some expanded perspectives on use of social science in public policy. In M. Yinger & S.J. Culter (Eds), *Major social issues: A multidisciplinary view* (pp. 346-357). New York, NY: Free Press.

Perkins & Will. (2013, December). *Research.* Retrieved from http://perkinswill.com/research

Popov, L. (2009). Modes and models of environment and behavior research utilization: Toward analysis of design-research interactions. Proceedings of EDR 40. P 342.

Phares, E. (2011). The state of evidence-based design in healthcare interior design practice: A study of perceptions, use, and motivation. (Unpublished master's thesis). Florida State University, Tallahassee, FL.

Roth, S. (1999). The state of design research. *Design Issues, 15*(2), 18-26. doi: 10.2307/1511839

Salama, A. M. (2015). *Spatial design education*: New *directions for pedagogy in architecture and beyond.* Farnham, Surrey: Ashgate Publishing Limited.

Salama, A. M. (2008). A theory for integrating knowledge in architectural design education. *Archnet-IJAR: International Journal of Architectural Research 2*(1), 100-128. Retrieved from https://www.researchgate.net/publication/26503571_A_Theory_for_Integrating_Knowledge_in_Architectural_Design_Education

Seidel, A. D. (1985). What is success in E&B research utilization? *Environment and Behavior,* 17(1), 47-70. doi: 10.1177/0013916585171004

Sommer, R. (1997). Utilization issues in environment-behavior research. In *Toward the Integration of Theory, Methods, Research, and Utilization* (pp. 347-368). New York, NY: Springer US.

Steelcase. (2013, December). *Steelcase research overview.* Retrieved from http://360.steelcase.com/topics/design/

Thompson, J. A. (1992). Design research. In J. A. Thompson (Ed.), *ASID Professional Practice Manual* (pp. 47-50). New York, NY: Whitney Library of Design.

Wiess, C. (1979) The many meanings of research utilization. *Public Administration Review 39*(5), 426-431. doi: 10.2307/3109916

Wild, P. J., McMahon, C., Darlington, M., & Culley, S. (2010). A diary study of information needs and document usage in the engineering domain. *Design Studies, 31*(1), 46-73. doi: 10.1016/j.destud.2009.06.002

AUTHOR

Amy M. Huber
Assistant Professor, Interior Architecture + Design
Florida State University, U.S.A.
Email address: amattinglyhuber@fsu.edu

APPENDIX
Survey Instrument.

PART 1 Demographics

1. **What is your current title/position?**
   - Jr. Designer
   - Interior Designer
   - Senior Designer
   - Design Director
   - Project Manager
   - Principal/Owner
   - Other (*if other please describe*)

2. **In what market sector do you primarily practice?**
   - Commercial
   - Health & Wellness
   - Residential
   - Retail/hospitality
   - Education
   - Other

3. **What is your current age (years)?**
   20-30   31-50   51-65   over 65

4. **How many years of professional design-related experience do you have (years)?**
   0-1   1-5   5-10   10-20   over 20

5. **What is the highest level of education you have completed?**
   - No degree
   - Certificate (not-design related) *please list*
   - Certificate in a design related field
   - Associate's degree (not design related) *please list*
   - Associate's degree in a design related field
   - Bachelor's degree (not design related) *please list*
   - Bachelor's degree in a design related field
   - Master's degree (not design related) *please list*
   - Master's degree in a design related field
   - Ph.D. (not design related) *please list*
   - Ph.D. in a design related field

6. **How large is your firm in terms of size of design staff?**
   - Sole Proprietorship
   - 2-5 Designer
   - 6-20 Designers
   - 21-50 Designers
   - 51-100 Designers
   - Over 100 Designer

PART 2 Research Activities & Preferences

This research uses a definition of design research as the following....
   "the identification of important design questions and the development and use of organized problem solving methods. It is a process for seeking and finding answers."   Thompson, 1992

7. **Using the above definition do you currently conduct design research to improve your design solutions?**
   Yes or   No

If the answer is Yes to item 7...
8. **What type of research do you take part in? (Mark all that apply)**
   - Client research
     Examining a client's business plan, strategic goals and/or lifestyles to inform programming
   - Interviews/Focus Groups
     Meeting with key stakeholders to better understand project requirements
   - Observation
     Observing patterns, preferences, workflows, etc. of clients or end-users while in their current space
   - Precedent Studies
     Systematic analysis of previous projects
   - Human Behavior & Theory
     Exploring relevant psychological, behavioral, or sociological phenomena
   - Business/Lifestyle Trends
     Exploring broad topics that may influence your designs
   - Design Trends
     Understanding what is being done on other projects by competitors
   - Sustainability Issues
     Exploring better systems, methods, or product specifications with a goal of sustainable outcomes
   - Product Research/Prototyping
     Reading product literature, building or fabricating mockups for projects
   - Post Occupancy Evaluations
     Formalized research regarding project successes and failures after project completion and move in
   - Other (*if other please describe*)

9. What types of information are accessed?
        (Select all that apply)
                Academic Journals
                        if selected which
                Professional organization knowledge sites (e.g. IIDA Knowledge Network, ASID Knowledge Center)
                        if selected which
                Trade Publications
                        if selected which
                Product Literature
                Attend conferences/CEUs
                Discussing topics with colleagues
                Informedesign
                Other

        For those that don't select Academic Journals

10. What is the primary reason you don't use Academic Journals?
                I don't know about them
                I don't know how to access them
                I don't have access to them
                I don't have time for them
                I don't understand them
                Other? Please explain

11. How much time would you dedicate to reading one specific article?
        Less than 2 minutes   2-10 minutes   10-20 minutes   more than 20 minutes

12. What attracts you to a specific article? (Mark all that apply)
                The topic
                The graphics
                The author
                The source in which it is published
                Other

13. After completing the reading rank what you are most likely to remember the next day? Likert
                details such as statistics
                big ideas/conclusion
                specific headings
                specific images and accompanying graphics
                stories behind the research-what led researchers to conduct the study, how it was conducted etc.

14. Are there any ideas you would like to share regarding how designers may use or receive research findings?
        Open-Ended

15. Is there anything else you'd like to share regarding your research styles/preferences?
        Open-Ended

Thank you for your participation

(Exit)

8

AN EVALUATION OF STAIRWAY DESIGNS FEATURED IN ARCHITECTURAL RECORD BETWEEN 2000 AND 2012

Karen Kim, Edward Steinfeld
Center for Inclusive Design and Environmental Access,
University at Buffalo, The State University of New York, Buffalo, NY, USA

*Corresponding Author's email address: kskim3@buffalo.edu

···

Abstract

This paper discusses an evaluation of stairway designs featured in Architectural Record, a leading architectural professional journal, over a thirteen-year publication period (2000 to 2012). Images of stairways were classified as either hazard-free or hazard(s)-present using a hazard identification checklist, and the frequency of visible design hazards was tabulated. A total of 578 stairways were scanned in articles and advertisements, of which 78 (13.5%) were product advertisements. Sixty-one percent of the stairways had at least one visible design hazard including nearly half (47%) in product advertisements. The three most common hazards in stairways were inadequate handrails (161, 27.8%), excessive length of stairway flights (74, 12.8%), and low visual contrast on tread edges (73, 12.6%). The high prevalence of stairway design hazards in the professional literature indicates a need for improved professional education and media attention to safe stairway design.

Keywords: *Architecture; Stairway Safety; Evaluation Research; Hazard Identification Checklist; Environmental Design; Design Education.*

INTRODUCTION

Designing stairways is a ubiquitous part of architectural design. By today's standards, stairways should be designed and constructed to prioritize safety and usability for people of all ages and abilities. In the U.S., stairway trips, slips, and falls result in nearly 1,900 deaths (NSC, 2011) and 1,300,000 hospital emergency room visits per year (Pauls 2011a). The high incidence of stairway accidents is common not only in the US but also in many other countries including Canada (Pauls, 2011b), the U.K. (Roys, 2011a), Japan, and Sweden (Templer, 1992; Scott, 2005). In all these countries, building regulations include many requirements for safe design and construction of stairways yet there are still a large number of accidents. What are the causes of this public health problem? Can architectural research do anything about it?

Causes of stairway falls include risky behaviors, poor maintenance, and design failings (Templer, 1992). Risky behaviors include running, using electronic devices on stairways, and carrying things that obscure one's view or change the dynamics of balance. Maintenance causes include defective stairway features (e.g. loose treads, broken lighting), unsafe materials on the tread surface (e.g. ice, worn surfaces), and poorly conceived countermeasures intended to reduce falls (e.g., peeling of applied non-slip surfaces). Behavioral causes can be reduced by raising awareness of the risks of using stairways and thus increasing user caution, but experts argue that such actions may not be sufficient to mitigate the risks posed by design and construction of stairways since these risks are often not noticeable to users (Roys, 2001). Examples include ungraspable handrails and irregular step geometry.

A reasonably safe and usable stairway is defined in the literature as one that meets safety standards for three basic criteria: step geometry, handrails, and stairway visibility (Pauls, 2013). Step geometry should be uniform in shape and dimension and facilitate gait (Novak et al., 2016; Pauls & Barkow, 2013; Johnson & Pauls, 2010; Jackson & Cohen, 1995); handrails should be both reachable and graspable (Maki, 2011; Dusenberry et al., 2009), and stairway components, i.e., steps, handrails, landing and headroom, should be clearly visible and perceivable to users (Archea et al., 1979; Sloan, 2011). These basic features are even more important for people in need of additional stair climbing support, i.e., people with physical, sensory, or cognitive limitations. Incorporation of these features leads to a universal design approach that would be safer for a wide range of people (Steinfeld & Maisel, 2012; Pauls, 2012).

Stairway design plays an important role in health promotion due to the health benefits of stair climbing as a form of exercise - improved weight control, lowered cholesterol levels, and improved cardiovascular fitness (Lee et al., 2012; Lewis & Eves, 2001). Research has demonstrated that building design can promote more frequent bouts of walking (Boutelle et. al., 2001; Nicoll, 2007; City of New York, 2010), but stairways are believed to have a higher potential for increasing light to moderate physical activity in part due to their presence and potential for use in every multistory building (Mansi et al., 2009; Cohen, 2013). This has led to the development of policies to improve the appeal of stair climbing, particularly in the U.S. where more than 66% of adults are obese (Brown et al., 2009). Policies include New York City Mayor Michael Bloomberg's executive order for architects to use "Active Design" strategies by designing highly visible, easy to access, and attractive stairways in new and renovated city buildings (City of New York, 2013). However, experts have also cautioned that increased stairway use could increase the number of stair-related injuries (Pauls, 2012), thus architects and builders need to pay careful attention to the design details and construction of stairways as they encourage use of stairways to improve fitness.

Despite some advances in knowledge of stairway safety, potentially hazardous stairway design practices seem to be prevalent. This suggests that the knowledge available on design of stairways is not being utilized. A scan of stairway images across a broad spectrum of media, including popular professional journals and internet blogs, will uncover many recently constructed stairways with identifiable and well known safety hazards that increase a person's risk of tripping, slipping, or falling. The stairways found in the media often have features that clearly do not meet safety standards, yet they were somehow not only constructed but also highlighted in feature articles and websites as exemplars of architectural design practice. Building codes in the U.S., including the International Building Code (IBC), American National Standards Institute (ANSI), and ADA Accessibility Guidelines (ADAAG), have made stairways in newly constructed public buildings safer by requiring architects to design stairways that are part of a means of egress to meet highly technical requirements; however, the criteria for stairways that are not part of a means of egress or in private dwellings are less stringent. With few exceptions, the IBC requires at least two general means of egress and not less than one that is "accessible" in buildings (ICC, n.d.), which means that stairways that are not part of a means of egress in any multistory building could have design features that would be considered hazardous and still comply with building regulations. Many unusual features are being incorporated into stairways such as glass stair treads, interactive sound and light, treads at acute and obtuse angles, etc. that are not addressed by regulations at all. These stairways are often centerpieces of the design, or "feature stairways," the most visible and most likely to be used by building inhabitants and visitors. The implications of these contemporary practices on safety are currently unknown.

We conducted a literature scan of stairways using *Architectural Record*, a leading professional architectural journal to investigate trends in stairway design practices. Professional journals are important to the field of architecture because they are a source of contemporary design ideas, product reviews, and continuing education for professionals and are often referenced during the research phase of projects (Borg & Gall, 1989 cited in Waugh, 2004). The purpose of this study was to examine current practices in stairway design as featured in the architectural press, assess the degree to which safe design practices are present, and identify issues that have not yet been addressed in stairway research. The results of this study suggest that safe stairway design practices should be supported by improved professional education and more media coverage on this topic. Several knowledge needs were also identified for further research in this field.

METHODS

Sample
We evaluated images of constructed stairways that were published in *Architectural Record* articles and advertisements between 2000 and 2012. *Architectural Record* was chosen because it is the oldest and most established professional architectural journal in the U.S. with a circulation of 115,155 (Ulrichsweb, 2014). This journal is considered an essential resource to architectural education and is included on the Association of Architecture School Librarians (AASL) Core List of 53 periodicals (2009 edition) (AASL, n.d.) – a list that is used as an evaluative criterion in the process of accrediting architecture degree programs by the National Architectural Accrediting Board (NAAB). Furthermore, *Architectural Record* has an h-index of 4 in the citation database produced by Scopus (SJR, 2014), which is the highest rating of any trade journal in architecture on the list (see Table 1).

Table 1: H-index of professional architecture journals on the AASL Core List covered in the Scopus database (Source: Authors).

Professional architecture journal*	h-index
Architect	2
Architectural Record	4
Architectural Review	1
Landscape Architecture	3
Lotus International	1
Planning	6
Preservation	1

*The Scopus database does not cover all titles on the AASL Core List. Note that *Planning* is the journal of the American Planning Association and does not have significant content related to building design.

Design

The study began as a class project by twenty-one graduate students in an ergonomics course at the University at Buffalo Department of Architecture. The class was divided into teams and assigned to scan the literature of different publication years. Each team developed a unique method of collecting and analyzing the stairways, including use of rating scales and checklists. Consequently the results of the project varied, but each team identified a significant number of stairways with safety hazards that provided insight into contemporary stairway design practices. These findings indicated that a more controlled study would be fruitful for identifying a gap in the application of research knowledge to practice. The authors used the student work to develop a new method to quickly identify common stairway design hazards, referring to the literature on stairway safety to validate the items on the list. A hazard identification checklist was organized into four categories: railings, steps, visibility, and other (see Figure 1). A systematic review was then conducted by a single researcher (the first author).

Railing	
Handrail(s) not fully extended at top/bottom of flights	
Missing/inadequate balustrade(s)	
Missing/inadequate handrail(s)	
Handrails too large/too thin	
Steps	
High/low riser-to-tread ratio	
Irregular riser height and/or tread size	
Narrow stairway width	
Short tread depth	
Visibility	
Low visual contrast on tread edges	
Open risers	
Poor stairway lighting	
Distracting pattern on steps	
Other	
Excessive length of stairway flight	
Inconsistency within the top/bottom steps	
Oblique stairway shape	
Obstruction on stairway	

Figure 1. Stairway Design Hazard Checklist (Source: Authors).

Procedure

Every page in each issue of the journal was manually reviewed for images of stairways constructed in the U.S. Images that were readily discernible were documented using a scanner and/or digital camera. Small prints lacking sufficient details, and stairways located outside of the U.S. were excluded. Each image was cropped and inserted into a page template using graphic representation software. Evaluations were based solely on image content and guided by two principles: if a stairway image showed at least one condition listed in the hazard identification checklist, then it was classified as unsafe and 'hazard(s)-present'; if the image did not show any of the conditions listed, then it was considered reasonably safe and classified as 'hazard-free'. For each stairway, information on conditions observed, setting (public or residential) and image type (article or advertisement) was recorded on the checklist in spreadsheets, and the frequency of each condition on the list was tabulated. Some of the conditions in the checklist require precise measurements to ascertain their presence if deviations are only slight, e.g., high/low riser-to-tread ratio, irregular riser height and/or tread size. In this study, we only could identify obvious evidence of such conditions. Thus the results clearly understate the presence of unsafe conditions but a conservative estimate of frequency of problems is sufficient to achieve the goals of the research.

RESULTS

A total of 578 stairways were scanned over a thirteen year publication period between 2000 and 2012—of these, 78 (13.5%) were in product advertisements. The majority of the stairways in our sample were located in public settings (72.8%, n = 421).

Sixty-one percent of the total sample of stairways (n = 355) had at least one obvious design hazard and were classified as 'hazard(s)-present'—thus less than 40% of stairways in this study were considered reasonably safe (n = 223). Of those classified as hazardous, 62% were public (n = 219) and 38% were residential (n = 136) (see Figure 2). The results of the evaluation are presented in Tables 2 and 3. In advertisements, nearly half of the stairway products exhibited hazards (47%, n = 37) (see Figure 3).

The most frequently observed hazard category was railings, comprising 36% of all hazards documented. Visibility was the second most frequently observed hazard category (29.1%), followed by other hazards (19%), and steps (16%).

The three most common design hazards were missing or inadequate handrails (27.9%), excessive length of stairway flights (12.8%), and low visual contrast on tread edges (12.6%) (see Figure 4). While the vast majority of stairways showed only one (48%), two (32%), or three (15%) hazards out of sixteen that were included in this study, the number of hazards should not be used to rate stairway safety since a misstep or a fall can be caused by even one condition. Moreover, grievous and obvious conditions, like a steep stairway, with a long flight, lacking both handrails and balustrades, could be so obvious that users adopt a more cautious and attentive behaviour while using it or avoid using it. Evaluating the severity of problems on stairways in the sample was beyond the scope of this research.

The remainder of this section summarizes an evaluation of stairway design practices that were most commonly observed with references to the International Building Code (IBC)—the primary model building code adopted in the U.S. These building regulations are an indication of safety issues that are well known to governmental agencies and violations of design standards.

Missing or inadequate handrail(s)

Handrails serve multiple functions: visual cues to the stairway's presence, directional guidance, postural stability, fall mitigation, and reducing conflicts in ascent or descent by cueing stair users to stay to the side, usually to the right on stairways in North America (Templer, 1992; Jackson & Cohen, 1995; Dusenberry et al., 2009). Best practice recommendations include the provision of handrails on both sides of stairways. The International Building Code (IBC) has exceptions to the requirement for handrails on both sides of stairways. Notably, residential stairways and spiral stairways; decks and patios are not required to have handrails at all; a single elevation change at an egress door and changes in elevations of three or fewer risers in dwelling units also do not require handrails (ICC, 2011). Compliance only with minimum standards and taking advantage of exceptions in the codes can pose significant safety risks, especially for people who require additional support for balance, like children and elderly people.

Table 2. Public stairways: result of an evaluation of stairways featured in *Architectural Record* (n = 421). (Source: Authors)

Year	Hazard-free		Hazard(s) present	
	n	%	n	%
2000 (n = 35)	21	60.0%	14	40.0%
2001 (n = 30)	17	56.7%	13	43.3%
2002 (n = 40)	15	37.5%	25	62.5%
2003 (n = 37)	17	45.9%	20	54.1%
2004 (n = 36)	16	44.4%	20	55.6%
2005 (n = 26)	12	46.2%	14	53.8%
2006 (n = 39)	14	35.9%	25	64.1%
2007 (n = 26)	15	57.7%	11	42.3%
2008 (n = 27)	15	55.6%	12	44.4%
2009 (n = 38)	12	31.6%	26	68.4%
2010 (n = 21)	8	38.1%	13	61.9%
2011 (n = 29)	16	55.2%	13	44.8%
2012 (n = 37)	23	62.2%	14	37.8%

Table 3. Residential stairways: result of an evaluation of stairways featured in *Architectural Record* (n = 157). (Source: Authors).

Year	Hazard-free		Hazard(s) present	
	n	%	n	%
2000 (n = 15)	1	6.7%	14	93.3%
2001 (n = 14)	1	7.1%	13	92.9%
2002 (n = 11)	0	0.0%	11	100.0%
2003 (n = 13)	3	23.1%	10	76.9%
2004 (n = 12)	3	25.0%	9	75.0%
2005 (n = 13)	1	7.7%	12	92.3%
2006 (n = 13)	3	23.1%	10	76.9%
2007 (n = 18)	1	5.6%	17	94.4%
2008 (n = 22)	4	18.2%	18	81.8%
2009 (n = 10)	2	20.0%	8	80.0%
2010 (n = 5)	0	0.0%	5	100.0%
2011 (n = 5)	2	40.0%	3	60.0%
2012 (n = 6)	0	0.0%	6	100.0%

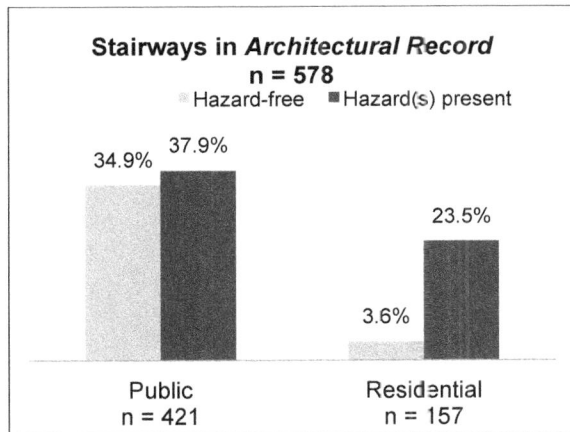

Stairways in *Architectural Record*
n = 578
Hazard-free   Hazard(s) present

34.9%   37.9%
23.5%
3.6%

Public
n = 421
Residential
n = 157

Figure 2. An evaluation of stairways featured in *Architectural Record* by setting. (Source: Authors).

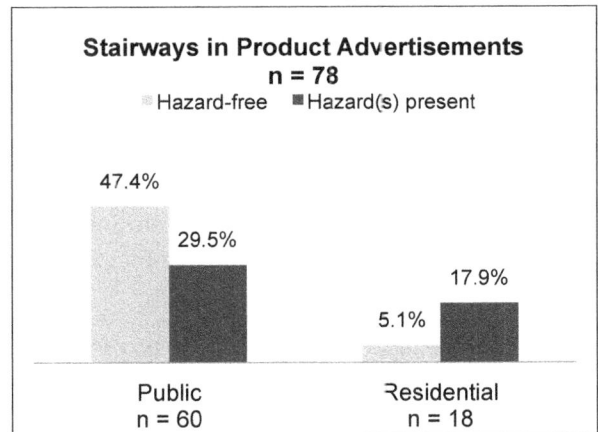

Stairways in Product Advertisements
n = 78
Hazard-free   Hazard(s) present

47.4%
29.5%
17.9%
5.1%

Public
n = 60
Residential
n = 18

Figure 3. An evaluation of stairways featured in *Architectural Record* product advertisements by setting. (Source: Authors).

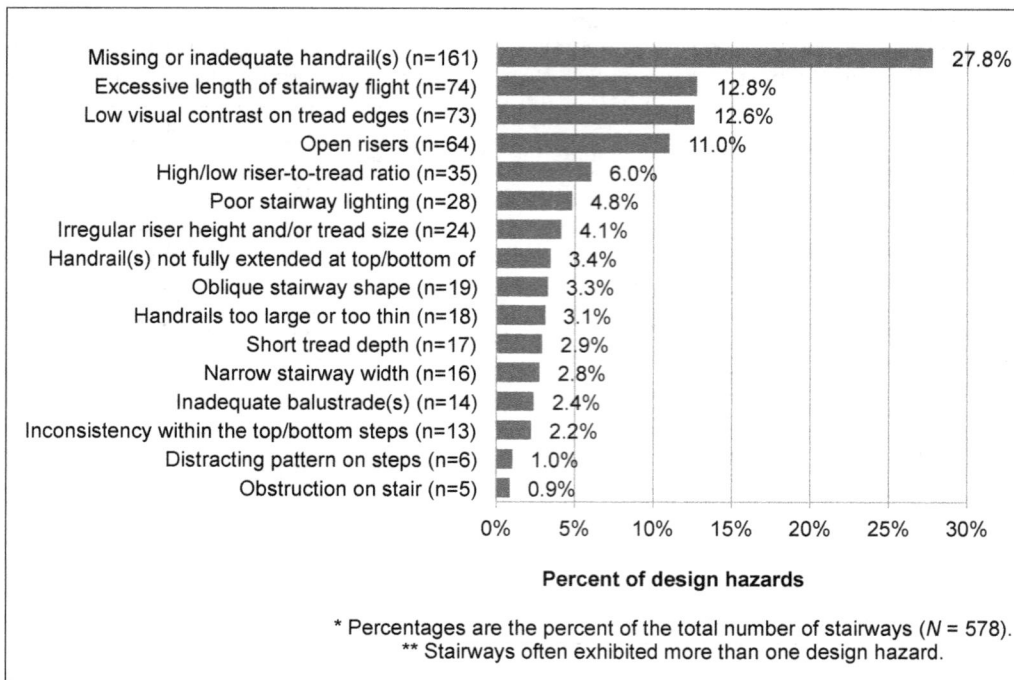

Figure 4. Frequency of stairway design hazards in *Architectural Record* feature articles and product advertisements between 2000 and 2012 (Source: Authors).

Although the handrail is probably the most important safety device for stair users, almost 30% of the stairways in this study (n = 161) had at least one or more of the following conditions: unprotected stairways at one or both sides (see Figure 5), three or fewer stairs that were potentially difficult to see or expect without handrails as visual cues, wide stairways without intermediate handrails distributed across the width, and non-continuous handrails that were interrupted by newel posts or other objects in environments where releasing grip of the handrail poses a risk.

Handrail(s) not fully extended
A basic principle of safe stairway design is that the handrail must be available for the user to grasp on the first step and maintain a grip all the way to the last step. Handrails are most heavily used at the top and bottom of flights; 70% of stairway accidents occur at these locations, demonstrating their importance (Templer, 1992). This research also suggests that handrails are needed most at landing areas. Handrail extensions or continuous handrails on intermediate landings can help people identify the start and end of the stairway, gain stability when mounting and dismounting flights, and make a safe transition in gait between landings (Danford & Tauke, 2001; Bakken et al., 2007). With proper installation, extended handrails can support people as they adapt their gait to various changes, e.g., in view, illumination, route direction, and floor surface (Templer, 1992) while entering and leaving stairways. U.S. building codes require handrails in a means of egress stairway to be continuous for the full length of each flight and extend at least 12 inches (305 mm) beyond the top and bottom riser, plus the width of one tread for the bottom extension. These extensions, however, are not required in dwelling units, assembly areas, the inside turn of stairways, and in existing stairways where they would be an obstruction (ICC, 2011).

In our study of new or substantially remodelled building projects, 20 (3.4%) stairways clearly had handrails of insufficient length, including those that truncated before even reaching the top or bottom riser (as seen in figure 5) and handrails that did not extend across the landing, primarily in public buildings.

Figure 5. Unprotected side, truncated handrail, low handrail visibility — duplex apartment in New York City, constructed 2008. (Photo courtesy of Jordi Miralles).

Handrails too large or too thin

After losing balance a stair user searches for support in an attempt to arrest a fall by reaching out and grasping a handrail (Templer, 1992). Handrail shapes that are too wide or too thin are not ergonomically designed to be grasped firmly and thus may not be effective during falls. A round handrail between 1.25 inches (30 mm) and 2 inches (50 mm) in diameter is generally accepted as the best shape and size for gripping because it provides a "power grip" in which the thumb can touch the fingers in the shape of the letter 'C' (Maki et al., 1998; Bakken et al., 2007; Dusenberry et al., 2009). Narrow or other shapes that require a "pinch grip" are generally not recommended (Maki, 2011).

Handrails were visibly too wide or too thin in 18 (3.1%) stairways. Rectangular shaped handrails made of metal or wood in 2x4 or 2x6 flat configuration were types of poor handrail shapes commonly observed. Railings with smaller cross-sections were used in minimalist designs, probably for their reduced obtrusiveness. As with irregularities of step geometry, the study methods did not allow measurement of handrails, and we could only see grievous conditions so this deficiency is probably underreported.

Inadequate balustrade(s)

Although balustrades are intended to protect sides of stairways, there is more leeway in regulations for creativity in baluster design than in railing design, and several types of hazards have been identified. Open spaces between balusters that are greater than 3.5 inches (90 mm) are areas where a child's head could slip through or body parts could become caught during a fall (Archea et al., 1979 cited in Templer, 1992). Large gaps under the bottom rail allow objects to slide or roll off stairways to areas below. Sharp edges of balusters pose risk of bodily injury. Horizontal rails and other balustrade attributes that can be climbed also pose risks for children and even adults (Templer, 1992).

The balustrade was missing or inadequate in 14 (2.4%) stairways. The majority of these (86%, n = 12) were located in private homes where the practice of omitting balusters was common. The findings provide evidence of a stylistic preference toward minimalist-aesthetic in stairway design. In the simplest of designs, only a railing was provided with no protective balustrade. Another variation on this theme was a railing with balusters spaced several stair treads apart and large enough for a person to easily fall through.

Low visual contrast on tread edges

The tread's leading edge is a key attribute in stairway design because it is crucial to help users perceive elevation changes between steps, place their foot accurately, and control their gait (Archea et al., 1979; Templer, 1992; Zietz & Hollands, 2009; Den Otter et al., 2011). Stair users visually scan a stairway using their foveal and peripheral visions. Depending on the user's attentiveness and the complexity of the environment, people may either look at the stair edges continuously, or they may glance at the stairs periodically. Most people scan the stairway at least once every seven steps taken (Templer, 1992), and rarely look directly at the stair edges. Miyasike-daSilva et al. (2012) suggest that this scanning behavior can be attributed to increased use of peripheral vision over foveal vision during the stair use task. Stair edges that are high in contrast levels can help make each step more visible in both central and peripheral zones of vision (Templer, 1992; Den Brinker et al., 2005).

U.S. building codes advise architects and builders to provide high visual contrast on stair tread edges mainly for people experiencing low vision (ICC, 2011), but this design practice can obviously benefit those with good vision as well.

In this study, low visibility of stair tread edges was observed in 73 (12.6%) stairways. This design practice was the third most commonly observed unsafe condition in stairway images. The results suggest that many stairways are clearly designed with a priority on aesthetics at the expense of safety. Stairways with a monochromatic color scheme for risers and treads are commonly used to achieve a cohesive visual form, uninterrupted by distracting elements (see Figure 6). Glass stairways and reflective tread materials contributed to the prevalence of this condition as well (see Figure 10).

Figure 6. Low visual contrast on tread edges, luminous lighting, obstacles on steps — Vera Wang Store in New York City, constructed 2010. (Photo courtesy of Paul Warchol Photography).

Open risers

There are two primary types of step and riser design: closed riser stairways (required within a means of egress in building regulations) and open riser stairways (not permitted within a means of egress by building regulations). Closed risers prevent feet and canes from accidentally slipping under treads, and they keep children, pets and objects on stairways from falling through (Templer, 1992). The solid barriers between treads also block distracting views in the background behind the stairway that may draw the user's attention away from the steps during ascent and cause tripping. Open riser stairways can cause a person to feel a sense of insecurity about the stair climbing task (Scott, 2005) and thus are not generally recommended. Where permitted, open riser stairways must be installed so that a 4 inch (100 mm) sphere cannot pass through openings (ICC, 2011).

In this study, we observed 64 (11%) open riser stairways with overstimulating surrounding views that would cause visual distractions (see Figure 7).

Figure 7. Views through open risers — Isabella Stewart Gardner Museum in Boston, constructed 2012. (Photo courtesy of Bruce T. Martin Photography).

Poor stairway lighting

Lighting affects our ability to perceive steps, railings and hazards in stairways (Templer, 1992). Research has demonstrated stairways with lower illumination levels have a higher incidence of falls. Carson et al. (1978) found that incidents were twice more likely at 2 foot-candles (22 lux) of light than 8 foot-candles (86 lux) (Templer, 1992). Hamel et al. (2005) found that older adults did not lift their legs as high off the steps while descending the stairs as young adults did, resulting in inaccurate stepping patterns, which makes them more vulnerable to missteps in low light conditions. Kasahara et al. (2007) found that under low illumination, older adults restricted their eye movements and visual scanning patterns to foveal regions because they required more time to focus on the steps directly ahead, and thus causes them to disregard visual information in the periphery.

Current lighting recommendations for stairways in the U.S. range from 10 to 20 foot-candles (108 to 215 lux) (Templer, 1992; IES, n.d.). Good stairway lighting should also include even illumination on the handrails and walking surfaces so that shadows do not fall on the stairway as well as indirect illumination that does not shine into the user's field of view and cause glare (Templer, 1992).

In this study, 28 (4.8%) stairways were noticeably dim or unevenly lit. Although it was impossible to actually measure the illumination in the photographs, the conditions we identified were very obvious without such measurements. Moreover, professional photographers carefully illuminate their subjects. Thus the frequency of this hazard is probably underreported.

High or low riser-to-tread ratio

The slope of the stairway should allow comfortable walking gaits to reduce the risk of falls (Novak et al., 2016). Sometimes, however, the stair pitch is skewed toward steeper slopes in order to fit the stairway in a building or to increase the economic efficiency of the building plan, i.e., more rentable or saleable space. Research shows there are more falls on stairways as steps depart from a "best practice" standard of a 7 inch riser (180 mm) and 11 inch tread (280 mm) (Templer, 1992). In particular, Johnson and Pauls (2010) demonstrated that stairways with high risers cause more falls. Ascending users require a higher leg lift and more strength to raise their body up and forward, and descending users shift their weight forward and downward for longer distances while balancing on one tread in each footfall (Templer, 1992).

The U.S. building code allows exceptions to the "7-11" design standard for residential and spiral stairways. In residences, riser heights can be a maximum of 7.5 inches (190 mm), and tread depths can start at 10 inches (255 mm). Spiral stairways can have higher risers measuring up to 9.5 inches (240 mm), and treads can be more narrow at 7.5 inches (190 mm) depth measured 12 inches (305 mm) from the narrower edge (ICC, 2011).

We could not identify stairways that depart from optimal conditions due to slight variations as they are not enough to be noticeable in photographs. We were only able to identify obviously excessively steep and shallow stairways and found that 35 (6%) stairways met the screening criteria, ranking fifth in stairway design hazards. Since we could not measure risers and treads, this result is most likely underrepresenting the frequency of stairways with non-optimal riser-to-tread ratios.

Irregular riser height and/or tread size

The riser and tread must be uniform for every step in a stairway flight. People tend to expect well built (and safe) stairways with uniformity of tread and riser dimensions throughout their length, an expectation that leads to low attention to the steps while climbing stairways (Templer, 1992). A continuous run of rectangular treads that are equal in shape and size can help reduce the risk of falls by allowing stair users to have a more consistent and natural gait as opposed to treads with varying sizes and shapes that force alterations of gait while climbing the stairway. It is well known that dimensional irregularity of steps is a leading cause of stairway falls (Jackson & Cohen, 1995; Cohen et al., 2009; Johnson & Pauls, 2010). To help counter this problem, the building code requires uniform risers and treads where the largest riser/tread minus the smallest riser/tread in a flight of stairs cannot exceed 3/8 inches (9 mm) (ICC, 2011). But, this tolerance may be too great. Research shows that even a slight irregularity of as little as 1/4 inch (6 mm) can interfere with the user's gait (Johnson, 2011). In construction, dimensional variations of between 1/6 to 1/4 inch (4 mm to 6 mm) are commonly observed (Roys, 2011b).

Twenty-four (4.1%) stairways in our sample had obvious irregularities in step geometry, including combinations of rectangular and winder treads. In Figure 8, the stairway is also located in a residential setting, in which residents are likely to pay even less attention to the changes in tread size than if they encountered it in an unfamiliar setting. As with other criteria, the study was limited in that we could not actually measure risers and treads so it is probably underrepresenting the frequency of irregular step dimensions. But, Figure 8 demonstrates that irregularity of shape is easy to observe from photographs.

Narrow stairway width

The clear width between walls, railings or the sides of stairways should accommodate the expected traffic flow and reach ranges for handrails (Templer, 1992; Levine, 2003). Adequate space on stairways is needed to move safely and comfortably, including a pacing, sensory and buffer zone for the user (Templer, 1992). The minimum code requirement for straight flight stairway widths is 36 inches (915 mm) for areas with an occupant load of 50 or less and 44 inches (1120 mm) for 50 or more people (ICC, 2011). Templer (1992) argues 38 inches (965 mm) is needed for minimal comfort; 56 inches (1420 mm) allows people to walk side-by-side in heavy clothing; but, a 69 inch (1755 mm) stairway width includes clearance between heavy clothing and tolerance for tracking error and thus is most comfortable to the user. Codes allow spiral stairways to be narrower, 26 inches (660 mm) in width (ICC, 2011), but these stairways do not provide the standard 11 inch (280 mm) minimum tread depth at the inside walking line; the minimum stairway width would have to be 6 feet, 9 inches (2.06 m) wide to provide adequate winder tread depth at the walking line (Templer, 1992).

In this study, 16 (2.8%) stairways were very narrow. It was noted that many of these also lacked handrails or had winder treads. Narrow stairways without handrails can create problems in implementing handrail retrofits in the future since handrails take up at least 3 inches (75 mm) on each side of stairways (Templer, 1992), and this would reduce the effective width of the stairway even more. A narrow stairway width in winding configurations forces the user closer to the inside radius where the tread becomes too small for safe walking (see Figure 8).

Figure 8. Irregular tread size, narrow stairway width, short tread depth — duplex apartment in New York City, constructed 1999. (Photo courtesy of Michael Moran Photography).

Short tread depth

Slips due to overstepping treads in descent are the most frequent type of stairway falls (Bakken et al., 2007; Johnson & Pauls, 2010). The risk of overstepping increases where treads are too narrow to accommodate the length of the foot. This condition is often found along the inner radius of winder or "pie-shaped" treads on stairways that turn or spiral. U.S. building codes require an 11 inch (280 mm) tread as the minimum effective depth for accessible stairways in public buildings (ICC, 2011). Treads on residential stairways are currently allowed to be smaller, with a minimum of 10 inches (255 mm), and on spiral stairways, treads are allowed to be even smaller with a 7.5 inch (190 mm) minimum depth measured 12 inches (305 mm) from the narrower edge (ICC, 2011). Although the study methods did not allow us to measure the steps, we found 17 (2.9%) stairways had treads that were obviously too narrow for proper foot placement (see Figure 8).

Excessive length of stairway flight

A reasonably safe stairway should consist of at least three risers so that it is noticeable and people do not accidentally walk into it (Templer, 1992). But, it should also not have too many steps without a landing since the risk of falling on stairways increases with longer durations of exposure. Moreover, excessively long stairway flights require greater energy expenditure over a longer time of exertion, which can lead to a sudden loss of balance for older adults, people with arthritis and those with low stamina who need to stop and rest periodically. There are psychological factors to consider as well. Using stairways can be a daunting task for people with physical or cognitive limitations. A long continuous run of steps can contribute to the fear of falling, especially among elderly people (Tiedemann et al., 2007). To encourage more stairway use for fitness, stairways should not induce avoidance behavior by appearing too challenging and dangerous.

Although the IBC (2012 edition) does not specify the number of steps in a flight of stairs, it is recommended by the National Safety Council (NSC) that a landing be provided at every tenth or twelfth tread (Reese, 2009). Older people in a focus group study reported that they can negotiate twelve steps maximum in between landings (Ormerod, 2011). In our study, we applied a tolerance of three steps. Thus stairway flights with more than 15 steps were identified as hazardous, including the stairway with twenty-two risers shown in Figure 9. An excessive length of stairway flight was the second most common design hazard (12.8%, n = 74).

Figure 9. Excessive length of stairway flight, low visual contrast on tread edges, inadequate balustrades —DVF Studio in New York City, constructed 2007. (Photo courtesy of Elizabeth Felicella).

DISCUSSION

Our scan of current stairway design practices suggests that the knowledge available on stairway safety is not being applied consistently in buildings featured in a leading trade journal. Although standards and codes lag behind the science, that does not excuse the adoption of best practices, for example, contrasting tread edges and easy to grasp handrails, that have been well known for decades.

Architectural Record, like most professional design journals, features buildings that have unusual formal characteristics, including stairways, presumably to attract readership and maintain an innovative edge in the field of architectural journalism. The stairways featured are not the prosaic stair towers hidden away in the bowels of buildings, nor those in the buildings with which most citizens are familiar, e.g., most stairways in schools, formula driven office buildings, health facilities, and spec houses. One could argue that the vast majority of stairways in buildings meet code requirements and are relatively safe. Yet our study demonstrated that, in contrast, most stairways selected for publication are potentially dangerous based on well-known causes of stairway accidents. The results suggest that many architects are either unaware of the risks associated with stairway use, or they choose to ignore good practices to achieve other goals, such as getting their work published, attracting clients, or recognition by their peers (e.g., creating the lightest looking stairway ever). The higher value placed on form is not confined only to those architects whose work is featured in the architectural press. From our own personal experience, as a recent student and a design instructor who has attended hundreds of studio reviews, many design studio instructors encourage students to make stairways unusual and even scary! The focus on form over function can be perpetuated in design studio courses by failing to pose a design problem that requires a design solution (Maturana, 2014). Some clients clearly agree with this value orientation, or such stairways would not be approved for construction. They may even urge their architects to create a "wow" experience. But, it is the architect's responsibility to ensure that buildings are safe and resist such pressure.

We selected *Architectural Record* for the source of examples because it is currently the most established, well respected, and well-read professional architecture journal in the U.S. and thus serves as an "opinion leader" in the professional print media. Interestingly, we found only two feature articles about stairway safety in our scan of thirteen years of issues (Talarico et al., 2000; Arsenault, 2012), suggesting communication about safe stairway design is limited in industry news and trade magazines in the print media. Considering stairways are important parts of architectural design and given the liability risks associated with their use, it would be reasonable for professional journals such as *Architectural Record* to feature more articles about good stairway design practices or at least provide some criticism of the unsafe stairways featured in their journal. These articles could also focus on the health benefits that stairways provide building users. Such content could influence architects and builders to design and build more carefully.

We found that 61% of the stairways had at least one visible and well-known design hazard, including almost half (47%) of the stairways in product advertisements. Although the prevalence was lower in advertisements, the result raises questions about company standards for product safety, e.g., how well-informed are companies on their products' safety features?

The most common design hazards were defective or missing handrails, long stairway flights, reduced visibility of tread edges, open risers, and non-optimal riser-to-tread ratios. The results are aligned with the top priority-issues that have been identified in stair research and outlined by Pauls (2011c): handrail and guard shapes, step geometry, tread characteristics, landing size, stairway widths, and factors related to visual perception and cognition. We also found a much higher proportion of unsafe stairways in homes (87%) in comparison to those in public buildings (52%), which alludes to the "greater need" of improving stairway safety in homes – the site of 90% of falls (Pauls, 2013). Unsafe behaviour is more likely to occur in a residence due to the presence of children and the familiarity of the setting. Yet, residential stairways are typically less regulated, and so given the leeway, it seems architects experiment even more in these settings and are clearly given client support.

The results suggest a need for improved professional education on stairway safety for architectural journalists, in continuing education, and in the academy. While this study cannot be used to attribute negligence to the majority of practicing architects, it does suggest that opinion leaders in the profession place a higher priority on aesthetics than on safety and health or are lacking in knowledge about this important aspect of design. To counter the pervasive impact of media, more attention to building codes and design standards, including their limitations with respect to research knowledge, should be incorporated into school curricula to ensure that students learn best practices. Moreover, attention should be given to improve and expand stairway safety requirements in building regulations to protect the unsuspecting public, including housing. Public educational programs should also include emerging issues like the impact of obesity on stairway falls (Pauls, 2012), reduction of distracted walking accidents caused by use of electronic devices (Caya, 2014), and effects of new technologies and design concepts, like glass stair treads and minimalist handrails, on user safety. A review of the legal practices surrounding stairway accidents and risks of legal action against architects, building owners, and product manufacturers should be an important topic for continuing education and professional practice courses. In particular, due to gaps in the regulations, professionals and students need to learn best practices so that they are aware that standards and codes lag behind the research knowledge. Best practice examples would not only inform professionals, but also encourage interest in building safer stairways.

Our systematic research raised important issues for future investigation. First, open risers were of particular interest to us because the use of this feature is almost entirely based on aesthetic considerations. There is no practical reason for using open risers other than saving material, but the materials used on most open riser stairways, for example glass or stainless steel, clearly indicate that cost was not a major consideration in their design. In particular, understanding the impact of open riser stairways on both visual performance and gait is an area that needs research attention. Second, Templer suggested that long stairways are safer because they cause people to use more attention on the stairs (1992). In other words, in a short stairway flight, the user may glance at the steps only at the transition at top and bottom but not in between, but, on a long stairway, they will glance at the steps one or more times in addition. Is this hypothesis correct? Research on how users distribute their attention in relation to stairway length would be useful to determine optimum lengths of flights from an attention perspective. In addition, it would be useful to know the impact of long treads on users' perception of effort and benefits from stair walking. Do long stairways discourage use and therefore reduce opportunities for improving fitness? Third, winder treads are inherently more dangerous than straight stair treads, but they are used by architects to create sinuous stairways or to fit stairways in a small space. What is the impact of winder treads in places where environmental distractions cause people to turn their attention away from the stairs that are changing in size and shape as they descend or ascend? Finally, new technologies allow stairways to be constructed of unconventional materials. In this study, glass stairways were associated with low visibility of the tread edges (see Figure 10). Although nonslip treatments are available for glass used as a walking surface, where users are likely to track water in during bad weather, these treatments may not be adequate.

Research is needed to determine if glass tread edges can be easily perceived and whether non-slip surface treatments are adequate to prevent slipping, especially under wet conditions.

Figure 10. Open glass treads, irregular tread shape, missing handrails — duplex apartment in New York City, constructed 2001). (Photo courtesy of Roy Wright Photography).

There were several limitations to this study. The results represent stairway design practices in the U.S. Examples from buildings in other countries were not included because design standards differ across jurisdictions and countries. Thus findings cannot be generalized to buildings in other countries. Further, we only studied one professional journal. However, this journal covers noteworthy buildings and the work of leading architects. These same buildings are routinely featured in most other architecture trade journals. A comparative study could determine if other sources are doing a better or worse job in promoting safe stairway design. A comparison of Internet sources with print media would also be useful. Today the Internet may actually play a more important role in forming professional opinions. Like other traditional publications, *Architectural Record* itself maintains a major Internet presence, including a free site with limited content, additional in-depth material available to subscribers for a fee, and daily news sent by email. The content is similar to the print version although there are more buildings featured. Blogs about architecture like Dezeen and Architzer are proliferating and becoming more professional in their content over time. Searching websites and blogs is, of course, much easier than the print version and we intend on expanding our research to this material. We are anticipating that the trends we observed in this study will be similar or magnified even more on the Internet versions of publications and on blogs. For example, Dezeen has a prominent feature of "extreme staircases." We could not find one example in their collection of photos that would be considered a "safe" stairway by our criteria. This feature is actually focused on promoting the most extreme of stair design ideas at the expense of safety!

Another limitation of the study is that the buildings featured in media are not representative of the vast body of stairway construction. *Architectural Record* and other architectural trade journals focus primarily on high-end buildings, especially in the residential sector of the building industry. A similar study of buildings not included in journals would be useful to find out if the findings here are widespread in professional practice. The buildings featured are primarily those that are designed by well-known or established architects who may have more leeway to depart from conventional practice than the average architect. Other architects may be more knowledgeable and careful about safe stairway design. Yet, the architects whose work are featured in the journals are role models for the profession and one may ask whether they should take that responsibility seriously. They are like star athletes or performers whose fashions and lifestyles often set norms for their fans.

Although the study was carefully designed to maintain accuracy, the stairways were evaluated by a single researcher, thus the ability of other researchers to get the same results with the checklist is still not known. The method used also has some significant limitations on access to content. Stairway images could have been misleading or distorted during professional photography and editing; details could have also been hidden due to camera angles. But, given the simplicity of the method used and our conservative approach to evaluation, our findings are probably an underrepresentation of the problems found in current practices. With proper training in the use of the checklist and the evaluation task, it is also likely that good inter-rater reliability can be achieved using the checklist.

The checklist used in this study provides a tool for future research and practice. In future work, we also hope to extend the literature scan to learn more about the reader's perception of stairway designs. We also intend to assess the degree to which readers of professional journals can recognize unsafe stairway design features and whether, for the architects, it influences their own design practices, and for clients, their communications with design and construction professionals. And, through two other threads of research, in a laboratory setting and systematic observations of stairways in use, we are already taking up some of the gaps in our knowledge about safe stairway design that we observed in this literature scan.

In conclusion, this research raises questions about professional values that deserve more research. Is valuing form over function common among the opinion leaders in the architectural profession? Could stairway safety be only one example of this value orientation? Such a value orientation may be an occupational hazard – a by-product of developing strong design values and getting noticed. But, if it contributes to mistrust of the profession and the need for increased vigilance on the part of clients and building regulatory officials, it should be addressed in professional education and in the professional media. Most importantly, the unwary public is clearly put at risk without their knowledge when safety issues are neglected. How can such practices be changed? It is a public health question that the profession, including accrediting authorities, need to consider. We hope that with education and awareness, architects will take more interest in designing safer stairways that at the same time are attractive and innovative. These goals are not mutually exclusive.

ACKNOWLEDGEMENTS

The contents of this report were developed under a grant from the Department of Education, NIDRR grant number H133E100002. However, those contents do not necessarily represent the policy of the Department of Education, and you should not assume endorsement by the Federal Government.

REFERENCES

Archea, J., Collins, B. L., & Stahl, F. I. (1979). Guidelines for stair safety. Washington: U. S. Dept. of Commerce, National Bureau of Standards : for sale by the Supt. of Docs., U. S. Govt. Print. Off. Arsenault, P. J. (2012). Stair Treads and Nosing. Architectural Record, 200, 181-185.

Association of Architecture School Librarians [AASL] (n.d.). The AASL Core List of Periodicals. Retrieved July 7, 2014, from http://www. architecturelibrarians.org/page-1364618

Bakken, G. M., Cohen, H. H., & Abele, J. R. (2007). Slips, Trips, Missteps, and Their Consequences. Lawyers & Judges Publishing Company.

Borg, W. R., & Gall, M. D. (1989). Educational research: An introduction (5th ed). New York: Longman.

Boutelle, K. N., Jeffery, R. W., Murray, D. M., & Schmitz, M. K. H. (2001). Using Signs, Artwork, and Music to Promote Stair Use in a Public Building. Am J Public Health, 91(12), 2004-2006.

Brown, W. V., Fujioka, K., Wilson, P. W., & Woodworth, K. A. (2009). Obesity: why be concerned? Am J Med, 122(4 Suppl 1), S4-11.

Caya, C. (2014). Texting while walking injuring 'clueless' pedestrians. WBFO. Retrieved March 17, 2014, from http://news.wbfo.org/post/texting-while-walking-injuring-clueless-pedestrians

Carson, D. H., Archea, J., Margulis, S. T., & Carson, F. E. (1978). Safety on Stairs. U. S. Dept. of Commerce National Bureau of Standards: Washington.

The City of New York. (2010). Active Design Guidelines: Promoting Physical Activity and Health in Design. Retrieved January 14, 2013, from http://centerforactivedesign.org/dl/guidelines.pdf

The City of New York. (2013). Mayor Bloomberg Announces First Ever Center For Active Design To Promote Physical Activity And Health In Buildings And Public Spaces Through Building Code and Design Standard Changes. The Official Website of the City of New York. Retrieved August 5, 2013, from http://www1.nyc.gov/office-of-the-mayor/news/250-13/mayor-bloomberg-first-ever-center-active-design-promote-physical-activity-and

Cohen, J., LaRue, C. A., & Cohen, H. H. (2009). Stairway Falls. Professional Safety, 54(1), 27-32.

Cohen, S. M. (2013). Examining the effects of a health promotion intervention on the use of stairs. Journal of Articles in Support of the Null Hypothesis, 10(1), 17+.

Danford, G. S., Tauke, B., State University of New York at Buffalo. Center for Inclusive Design & Environmental Access., New York (N.Y.). Mayor's Office for People with Disabilities., New York (N.Y.). Dept. of Design and Construction., American Institute of Architects. New York Chapter., & New York (N.Y.). Office of the Mayor. (2001). Universal design New York. [New York, N.Y.]: City of New York Office of the Mayor.

den Brinker, B. P. L. M., Burgman, L. J., Hogervorst, S. M. J., Reehorst, S. E., Kromhout, S., & van der Windt, J. (2005). The effect of high-contrast marking of treads on the descent of stairways by low-vision people. International Congress Series, 1282, 502-506.

Den Otter, A. R., Hoogwerf, M., & Van Der Woude, L. H. (2011). The role of tread fixations in the visual control of stair walking. Gait & Posture, 34(2), 169-173.

Dusenberry, D. O., Simpson, H., & DelloRusso, S. J. (2009). Effect of handrail shape on graspability. Applied Ergonomics, 40(4), 657-669.

Hamel, K. A., Okita, N., Higginson, J. S., & Cavanagh, P. R. (2005). Foot Clearance during Stair Descent: Effects of Age and Illumination. Gait & Posture, 21(2), 135-140.

Illuminating Engineering Society [IES]. (n.d.). Light in Design - An Application Guide. Retrieved August 5, 2013, from http://www.iesna.org/PDF/Education/LightInDesign.pdf

International Code Council [ICC]. (2011). 2012 International Building Code (IBC). Country Club Hills, Ill: ICC.

International Code Council [ICC]. (n.d.). Accessible Means of Egress. January 14, 2013, from http://www.iccsafe.org/safety/Documents/MeansofEgressBroch.pdf

Jackson, P. L., & Cohen, H. H. (1995). An In-Depth Investigation of 40 Stairway Accidents and the Stair Safety Literature. Journal of Safety Research, 26(3), 151-159.

Johnson, D. A. & Pauls, J. (2010). Systemic stair step geometry defects, increased injuries, and public health plus regulatory responses. Contemporary Ergonomics and Human Factors 2010. London: CRC Press: 453-461.

Johnson, D. A. (2011). Consistency of Dimensions. Paper presented at the International Conference on Stairway Usability and Safety, Toronto, Canada, June 2011.

Kasahara, S., Okabe, S., Nakazato, N., & Ohno, Y. (2007). Eye Movement Patterns of the Elderly during Stair Descent: Effect of Illumination. Journal of Light & Visual Environment, 31(3), 134-140.

Lee, K. K., Perry, A. S., Wolf, S. A., Agarwal, R., Rosenblum, R., Fischer, S., . . . Silver, L. D. (2012). Promoting Routine Stair Use: Evaluating the Impact of a Stair Prompt Across Buildings. American Journal of Preventive Medicine, 42(2), 136-141.

Levine, D. R., State University of New York at Buffalo. Center for Inclusive Design and Environmental Access., New York (N.Y.). Mayor's Office for People with Disabilities., New York (N.Y.). Dept. of Design and Construction., & New York (N.Y.). Office of the Mayor. (2003). Universal Design New York 2. Buffalo, N.Y.: IDeA Publications, Center for Inclusive Design and Environmental Access [IDeA], University at Buffalo, The State University of New York.

Lewis, A., & Eves, F. (2001). Specific Effects of a Calorie-Based Intervention on Stair Climbing in Overweight Commuters. Annals of Behavioral Medicine, 42(2), 257-261.

Maturana, B. C. (2014). Where is the 'problem' in design studio: Purpose and significance of the design task. Archnet-IJAR, 8(3), 32-44.

Mansi, I. A., Mansi, N., Shaker, H., & Banks, D. (2009). Stair Design in the United States and Obesity: The Need for a Change. Southern Medical Journal, 102(6), 610-4.

Maki, B. E., Perry, S. D., & McIlroy, W. E. (1998). Efficacy of Handrails in Preventing Stairway Falls: A New Experimental Approach. Safety Science, 28(3), 189-206.

Maki, B. E. (2011). Handrails. Paper presented at the International Conference on Stairway Usability and Safety, Toronto, Canada, June 2011.

Miyasike-daSilva, V., & McIlroy, W. E. (2012). Does It Really Matter Where You Look When Walking on Stairs? Insights from a Dual-Task Study. PLoS ONE, 7(9), e44722.

National Safety Council [NSC]. (2011). Injury Facts, 2011 Edition. Retrieved January 14, 2013, from http://www.nsc.org/Documents/Injury_Facts/ Injury_Facts_2011_w.pdf

Nicoll, G. (2007). Spatial measures associated with stair use. Am J Health Promot., 21(4S), 346-52.

Novak, A., Komisar, V., Maki, B., & Fernie, G. (2016). Age-related differences in dynamic balance control during stair descent and effect of varying step geometry. Applied Ergonomics, 52, 275-284.

Ormerod, M. (2011). Persons with Disabilities and Older Users. Paper presented at the International Conference on Stairway Usability and Safety, Toronto, Canada, June 2011.

Pauls, J. (2011a). Injury Epidemiology. Paper presented at the International Conference on Stairway Usability and Safety, Toronto, Canada, June 2011.

Pauls, J. (2011b). Economics, Including Injury Cost. Paper presented at the International Conference on Stairway Usability and Safety, Toronto, Canada, June 2011.

Pauls, J. (2011c). Injurious Falls on Stairways: Background for international conference in June 2011. Proceedings of International Conference on Slips, Trips and Falls, Health and Safety Laboratory, UK, June 2011.

Pauls, J. (2012). Inclusive design of home stairways: Public health considerations and implementation challenges. Proceedings of 4th International Conference on Universal Design, Fukuoka, Japan.

Pauls, J. (2013). The pathology of everyday things – stairways – revisited. Proceedings of Annual Meeting of the Human Factors and Ergonomics Society, San Diego, USA.

Pauls, J., & Barkow, B. (2013). Combining risks from two leading factors in stair-related falls. Proceedings of the International Conference on Fall Prevention and Protection, Tokyo, Japan.

Reese, C. D. (2009). Industrial safety and health for infrastructure services. CRC Press: Boca Raton.

Roys, M. S. (2001). Serious Stair Injuries Can Be Prevented by Improved Stair Design. Applied Ergonomics, 32(2), 135-139.

Roys, M. S. (2011). Economics, including injury cost. Paper presented at the International Conference on Stairway Usability and Safety, Toronto, Canada, June 2011.

Roys, M. S. (2011). Consistency of step dimensions. Paper presented at the International Conference on Stairway Usability and Safety, Toronto, Canada, June 2011.

Talarico, W., & Nadel, B. A. (2000). Creating Safe and Appealing Public Stairs. Architectural Record, 188(6), 169.

Templer, J. A. (1992). The staircase : studies of hazards, falls, and safer design. MIT Press: Cambridge, Mass.

Tiedemann, A. C., Sherrington, C., & Lord, S. R. (2007). Physical and Psychological Factors Associated With Stair Negotiation Performance in Older People. The Journals of Gerontology Series A: Biological Sciences and Medical Sciences, 62(11), 1259-1265.

SCImago Journal & Country Rank [SJR]. (2014). SJR: Scientific Journal Rankings. Retrieved July 7, 2014, from http://www.scimagojr.com.

Scott, A. (2005). Falls on Stairways – Literature Review. Retrieved January 14, 2014, from http://www.hse.gov.uk/research/hsl_pdf/2005/hsl0510.pdf

Sloan, G. S. (2011). Perception & Cognition. Paper presented at the International Conference on Stairway Usability and Safety, Toronto, Canada, June 2011.

Steinfeld, E., & Maisel, J. (2012). Universal design: Creating inclusive environments. Hoboken, N.J.: John Wiley & Sons.

Ulrichsweb. (2014). ulrichsweb.com(TM) -- The Global Source for Periodicals. Retrieved July 7, 2014, from http://ulrichsweb.serialssolutions. com/title/1412613464691/42130

Waugh, C. K. (2004). The Core Periodicals in Career and Technical Education. Journal of Career and Technical Education, 20(2), 27-39.

Zietz, D., & Hollands, M. (2009). Gaze behavior of young and older adults during stair walking. Journal of Motor Behavior, 41(4), 357-365.

Karen Kim
Architectural Designer and Researcher
Center for Inclusive Design and Environmental Access, University at Buffalo
Email address: kskim3@buffalo.edu

Edward Steinfeld
Director of IDeA Center, Professor of Architecture, Architect
Center for Inclusive Design and Environmental Access, University at Buffalo
Email address: arced@buffalo.edu

9

URBAN AND RURAL UMAYYAD HOUSE ARCHITECTURE IN JORDAN: A COMPREHENSIVE TYPOLOGICAL ANALYSIS AT AL-HALLABAT

Naif A. Haddad[1], Fatima Y. Jalboosh [2], Leen A. Fakhoury [3], Romel Ghrayib [4]

1 Department of Conservation Science, Queen Rania's Faculty of Tourism and Heritage, Hashemite University, Jordan.
2 Private Practice, Amman, Jordan
3 School of Architecture and Built Environment, German Jordanian University, Jordan.
4 Department of Antiquities, Zarqa, Jordan.

Corresponding Author's email address: naifh@hu.edu.jo

Abstract

The Umayyad period represents one of the most prosperous periods in the history of Jordan. Most of the studies, however, have long been focused on palatial and luxurious architecture. In Jordan, few examples of Umayyad houses have survived in their entirety. However, the new discoveries at al-Hallabat rural houses allow an architectural enrichment of our knowledge for that period, even from a socio-economic point of view. In contrast with the better-known desert palaces that dominate the evidence for this period, they also assist in establishing the houses' typological patterns. This paper attempts to present and discuss the main Umayyad urban and rural house architecture in Jordan, while addressing al-Hallabat Umayyad houses based on recent unpublished reports and preliminary results of excavations. It aims to present a comparative typological pattern analysis of al-Hallabat houses excavated at two phases (1979-1982, 2002-2006) with parallel examples from Bilad al-Sham. The paper defines three typological patterns; nucleus, courtyard, and complex houses. All have at least one courtyard. The study shows that there were continuity and parallelism in Bilad al-Sham between these types and those used at least in early Byzantine and early Islamic period, such as these at ar-Risha and Khirbet al-Askar in Jordan.

Keywords: *Umayyad houses; nomad village; urban and rural settlement; architectural typology; courtyard house; al-Hallabat; desert palace*

INTRODUCTION: UMAYYAD JORDAN

The debated interpretation and explanation of the process of early Islamic settlement in Bilad al-Sham, based on recent archaeological evidence, is constantly growing, as more dynamic evidence comparing with literary references (Kennedy, 2014: 98). The buildings commissioned by the Umayyad dynasty in Bilad al-Sham, the core of the Umayyad realm illustrate the dynasty's appropriation and adaptation of Hellenistic, Roman, Byzantine, and Sassanian cultural traditions (Haddad, 2009:7; Arce, 2007; Almagro, 1992). Umayyad architecture in Jordan, actually, contains a mixture of eastern and western influences (Warren, 1978: 230; Haddad, 2009:1, 7). In fact, the early Islamic architectural elements were formed to respond effectively to people's physical, environmental, socio-economic and political, as also physiological and religious requirements at their time (Kaptan, 2013:5).

The Umayyad period represents one of the most prosperous periods in the history of Jordan due to its proximity to Damascus and its strategic geographic position which made it an important thoroughfare for pilgrims venturing to the holy Muslim sites in Arabia. Umayyad Jordan has also been the stage for great events that have influenced Islamic history and the Mediterranean region. The land of Jordan hosted the first confrontation between Islam and the Byzantine world in the battle of Mutah near Karak and the decisive battle of Yarmouk.

Perhaps one of the most important events from a political point of view was when the Abbasids launched the movement against the Umayyads using al-Humaymah in southern Jordan as their headquarters to establish their succession in Baghdad. In 750, Umayyad Jordan shifted to the rule of the Abbasids after the revolution that was initiated from al- Humaymah.

Jordan was enriched with some of the finest examples of early Islamic architecture, found anywhere including in caravan stops (caravanserais), bathhouses, and palaces at the eastern Jordanian desert. Interestingly, from a socio-economic point of view, the Umayyad period witnessed an expansion in urban and rural centers illustrated by the castles, the palaces, and the so-called 'Nomad Village', which stretched over great areas of Jordan. According to Kennedy (2014: 96) "there is a rarer site type — a dispersed village, known under various names", which he termed as 'Nomad Village', such as the ruins at Jabal Seys, at Qasr as-Swab, at Ar-Risha, at Hibabiya, and the those at Qasr el-Hallabat, which is the main subject of this paper. Kennedy (2014:107) states that 'Nomad Villages' "are important sites in their own right, revealing evidence for the progressive development of the pre-desert and adjacent desert regions of northern Jordan".

When the Umayyad inhabitants of Jordan were building these 'Nomad Village' complexes on the fringe of the desert/ Badiya, substantial Umayyad large urban towns existed at Jerash, Amman and Tabaqat Fahl (Pella) as also in many other long established towns such as Madaba, Hisban, Umm el Walid, Umm el-Jimal, Umm el-Rasas, and Aqaba (Alhasanat et al, 2012). In fact, the economics of many towns in early Umayyad Jordan became increasingly focused on the manufacture of tradable goods, especially in the eighth century (Walmsley, 2000: 305). In Jordan, the Umayyad achievements are reflected in the ability of the dynamic Muslim culture to expand far beyond urban centres to exploit in a creative management the reward of the agriculture and trade potential of formerly marginal frontier regions.

However, the al-Hallabat settlement provides the opportunity to investigate the rich cultural heritage significance of Umayyad domestic architecture, which is relatively limited in Jordan in comparison to palatial and formal architecture. The 'Nomad Village' settlement at ar-Risha (c. 5ha lies 165 km north-east of the Azraq Oasis and 35 km north of the small Baghdad Highway town of Ruwayshid) also has a collection of minimally preserved structures of individual buildings arranged in parallel lines with a mosque and large formal buildings.

The paper aims to present and discuss the main Umayyad urban and rural house architecture in Jordan, clarifying some socio-economic aspects while addressing al-Hallabat Umayyad houses based on recent unpublished reports and preliminary results of excavations. It attempts to present a comparative typological pattern analysis of al-Hallabat houses excavated at two phases, 1979-1982 and 2002-2006, with parallel examples from Bilad al-Sham.

BACKGROUND TO SECULAR UMAYYAD DESERT PALACE ARCHITECTURE AND PLANNING IN JORDAN: SOCIO-ECONOMIC AND POLITICAL INDICATIONS

Umayyad secular architecture, in fact, is best known from a group of desert palaces (often called Qasr in Arabic sources) constructed of stone and/ or brick in some cases. The so-called desert palaces have developed a unique architectural concept reflected in their location, their density, and their fast spread in a relatively short time (715-750) (Haddad, 2009: 2). Of significance were the events and activities of these early Islamic Umayyad palaces as reflected by their architectural typology while responding to the socio-economic aspects. Their remains were found mainly in the eastern desert of Jordan (Badiya), meanwhile, only a few were built in Syria (Qasr al-Hayr (727- 9), east and west) and a couple in the West Bank (Khirbat al-Mafjar in Palestine).

Such sites in Jordan are distinct from those in Syria; they are comparatively of modest scale and simple construction (Urice, 1987). It is not a mere coincidence that the greater part of the architecture attributed to Jordan corresponds to palaces or private residences and to the new oligarchy who sought to forgo a new image and mark the change of power, as Grabar (1987:

134-135) has pointed out. This peripheral and countryside category was normally the engine of politico-economic activities, and their economic, agricultural and technical innovations were intrinsically linked to urban centers and interregional networks.

These palaces and the other structures of the 'Nomad Villages' from the Umayyad golden age testify to Jordan's identity as a politico-economic center and as a major stop on the caravans' route. They demonstrate a face of the Umayyad life in the Middle East, which is not widely seen elsewhere, and some authenticate a perfect condition of preservation which is quite astonishing taking into account their vast epoch.

Recent results of GIS analysis (Alhasanat, et al, 2012: 343) show that these Umayyad palaces are carefully situated at the routes of transhumance and water sources. The distribution pattern of these prominent structures was strategically placed in the landscape to carefully monitor and protect the routes that led to Damascus. They clustered at the outlet of Wadi Sarhan, and there is, actually, line-of-sight communication between Azraq, Amra, Kharana, Muwaqqar, Umm al Walid, Mushatta, and Qastal. However, Qasr al-Hallabat and Qasr al-Tubah functioned more as two main patrol stations (Alhasanat, et al, 2012: 356-57).

These multi-functional activity structures were imposed by the nature of the emerging early Islamic state to strengthen the power and the economy of the newly established dynasty, but they also demonstrate how deeply the Umayyad culture had penetrated this provincial early Islamic area. Usually, these palaces are square in plan, with semi-circular towers buttressing the exterior walls; meanwhile the flanking entrance portals give the palace a fortified appearance (Fig.1a). The central square courtyard, generally surrounded by porticoes of two stories high, with the upper ground layout following the same guidelines as the lower. In fact, the square layout is not only conceived as a multi-functional space to control all the activities taking place, like trade at the Suqs, religious activities at the mosque, and political functions at the Qasr (Almagro and Arce, 2001: 665), but also as symbol of power, Just as a perfect balanced, stable, clear, and rigid form that reflects the concept of power and strength (Haddad, 2009: 6).

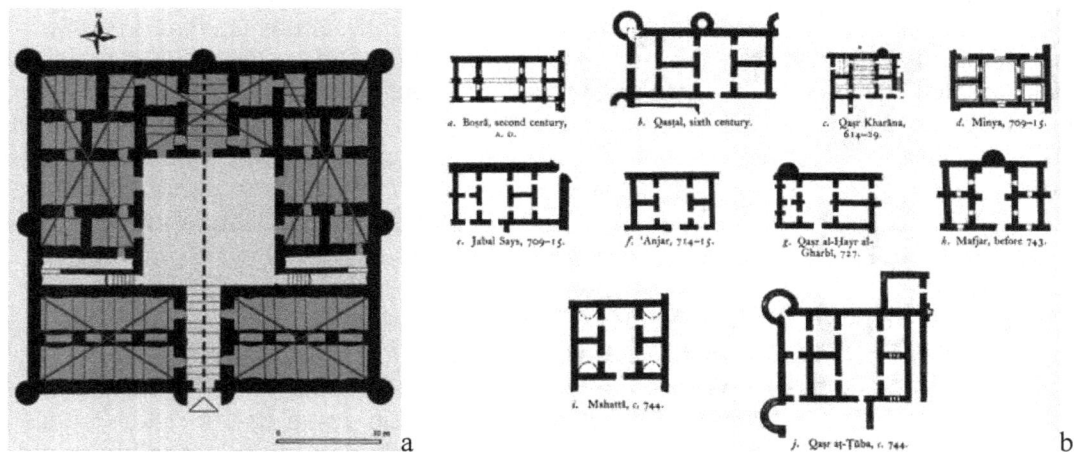

Figure 1. a. lan of Kharana Umayyad desert palace, as a model illustrating the interior symmetrical layout, space distribution, division and the bayts' units (After Haddad, 2009, Fig.5); b. different *bayt* units from Umayyad palaces in Bilad al-Sham (Source: After Creswell, 1989, Fig. 565).

The particularity of these buildings appears from the unity of the internal and external architectural form with the different variation of sizes and scales. They are characterized by clarity, identification, reflecting the image of the power of Islam from the outside, and the luxury from the inside reflecting the new lifestyle of the Umayyad (Haddad, 2009; 1, 7).

The Origin of the Architectural Pattern of the so-called Umayyad 'Bayt': Social Indications

In the early Umayyad architectural context, a bayt (plural buyut) is composed of a central hall flanked by a pair of rooms on either side from which the accessibility is achieved. This is a module frequently repeated in the desert palaces (Creswell, 1989: 516; Almagro, 1987:183; Haddad, 2009: 7) (see Fig. 1).

The bayt of the Umayyad palaces has different typologies that can be established from their architectural patterns; either independent or grouped structures, appearing in more or less compact ensembles. The independent type corresponds to buildings organized around a central square courtyard. Rooms open off the courtyard and are either directly or indirectly connected to it, such as at Qasr at-Tuba. These rooms form secondary spaces arranged around the main hall from which two or four adjacent rooms radiate. For example, at Kharana, other rooms were added to the three or five-room group (Fig. 1a), yet there is no repetition of any particular type of pattern from one case to another (Creswell, 1989).

Socially, this bayt arrangement is considered a more orderly expression of the same pattern seen in the few urban residences. According to Almagro (1992), these structures, which are based on the main hall and two to four smaller-sized adjacent rooms, appear to comprise the simplest type of room-unit, and can be compared to similar 'buyut', not usually found in the urban Umayyad house. Parallels, however, of such a module that resembles a bayt of the Umayyad places were found at the residential structures at Amman citadel, in both households of Building B and the main house over the Museum site (Harding, 1951: 7). This organization is clear in the room that may have functioned as a reception, opening onto two flanked rooms on either side of it, and thus forming a bayt (Fig. 2a).

It has been suggested that the architecture of these 'buyut' may reflect the Bedouin tent, Bayt al-Sha'ar. This theory was adopted by Helms (1990) through his study of the houses at ar-Risha, which was first suggested by Gaube (1979) in terms of a similar development of tents via a circular or square arrangement (dwar), as the design principle underlying the Qusur. However, Ahn (2010) clarified that one could identify the influence of the steppe and reference to the Bedouin tent by the normal Bedouin practice of laying out their tents in staggered rows facing downwind where each is separated from its identical neighbour by an acceptable distance. This scene is still reminiscent of many Bedouin towns today (Kennedy, 2014: 101; Ghrayib and Ronza, 2007: 423). However, only by this frame of Bedouin domestic forms, one can accept that it was derived from the tent, and in this sense, ar-Risha (see Fig. 9c) is a good example of what Bedouins might have built (Kennedy, 2014: 99, Fig. 4).

UMAYYAD HOUSES IN JORDAN:
KEY SOCIO-CULTURAL ASPECTS

While the most striking feature of Islamic architecture is the focus on interior space, the most typical expression of this feature is found in the inner space of the Muslim house. However, as the essential value in Islam is the emphasis on the inner aspect of self or thing (Bātin) and the subordination of the external aspect of self or a thing (Zāhir), the courtyard house and its organizational pattern are appropriate for the application of this principle (Hakim, 1986: 95-96). On the other hand, a comparative analysis of Islamic architecture in countries that have different climatic conditions suggests that climate has always been a significant force influencing the design and location of buildings (Toulan, 1980: 75). Therefore, the courtyard type house in Islamic architecture can be considered as a result of the integration of socio-cultural, religious, and climatic factors.

Historically, the courtyard type house, in fact, is a generic domestic form of residence that evolved independently in various ancient and traditional places. It is a product of cultural polygenesis dating at least to the Bronze age, and it has persisted in the Mediterranean area in the form of the classical atrium and *pastas* house to be adopted by Muslims in the dār al-Islam (Petruccioli, 2007:73). It is additive by nature; the severe and austere facades are presented to the outside world, and because of the darkness of the house interior, it provides secluded open space for all family and most domestic activities in the sunlit courtyard area (Ahn, 2010: 106).

It ensures privacy from outside or adjacent areas while providing a level of interdependence between neighbours with regard to the use and rights of shared walls, maintenance of streets, problems related to rain and waste water (Hakim, 1986: 95-96). It also allows the structure to expand with the growing extended families while it is easy to make additions to the original structures. This type can also be arranged as multi-smaller unit houses, containing several living units on one or more levels of the residence with the courtyard as a shared space (Ahn, 2010: 107).

Umayyad Houses within an Urban Context in Jordan
The purpose of this section is to realize if any typological patterns, wherever possible, have survived in the urban context in their entirety in Umayyad Jordan. This will be achieved by reviewing the basic house layout, its relation to the street, and the function of the household in relation to socio-economic conditions. Umayyad Jordan, Amman, Pella, and Jerash are the most representative of urban town centres. These three respective urban sites will be examined briefly before discussing al-Hallabat rural domestic settlement houses.

Jabal al-Qal'a (Citadel) Umayyad houses in Amman
The Umayyad palace complex, at Jabal al-Qal'a (citadel) in Amman, differs in its layout and architecture from the rest of the desert palaces in Jordan. From a political point of view, it was the administrative center and residence for the governor of the region. Still, the main area of the urban reform, undertaken by the Umayyads, was mainly the public space layout with a new urban concept to accommodate the organization of the newly created architectural elements and also the reuse of pre-existing features (Almagro and Arce, 2001: 662). It also included the construction of separate courtyard house units of a variety of sizes, ranging from two rooms and a courtyard to seven rooms, a latrine, and a courtyard. Meanwhile, the residential units of the palace in one structure have ten rooms, a latrine, a staircase, and a courtyard (Northedge, 1992: 157).

Excavations have uncovered a number of upper-class residences from the 7th to 8th centuries contributing to information on the socio-economic aspects. Although the sudden collapse of the buildings was attributed to the earthquake of 749 (Northedge, 1992: 142), a significant house (380 m^2) over the Museum site is preserved to a height of about 2.5m, built around a closed inner courtyard (Fig. 2a). The courtyard (8.6m wide) has a cistern with a shaft (Bennett and Northedge, 1976: 176). Plastered drains in the north-east and north-west corners of the courtyard conducted water from the roof to the cistern (Harding, 1951: 7). The cistern appears to have been constructed originally in the early Byzantine period.

Interesting also is the room that faced the courtyard, with the wide entrance. It was considered by the excavators to be a diwan (a reception room in the tradition of the Roman-Byzantine triclinium) (Bennett and Northedge, 1976). This possible reception room has a laid clay floor. The other lower-storey rooms apparently served as storerooms and workrooms. According to Harding (1951), parts of a mosaic floor were found on the upper storey, which apparently contained the living quarters.

However, Northedge (1992: 143) assumed that the building was apparently single-storey, as no evidence had survived of the roofing technique of a second storey or of a staircase to the roof. He speculated that the roof may have been barrel-vaulted, similarly to another building in the same area. The rectangular shape of the rooms would have accommodated barrel-vaulting, even at the expense of the regular thickness of the walls.

Tabaqat Fahl (Pella) Complex Houses

From a political point of view, Pella was an administrative district in the military province of Jordan in the early 7th century, serving the link between Damascus and Jerusalem: the two most important centres in southern Bilad al-Sham (Walmsley, 2008: 244; 1988: 144). However, the damage and the partial collapse of the domestic quarter of the main mound from an earthquake in 659-60 is evident, as indicated by the complete site destruction as well as from neighbouring sites.

This led to an urban modification translated by a rebuilding program that produced large houses and encroachment on public areas that continued until the end of the Umayyad period (Watson, 1992: 163-164; McNicoll et al, 1982).

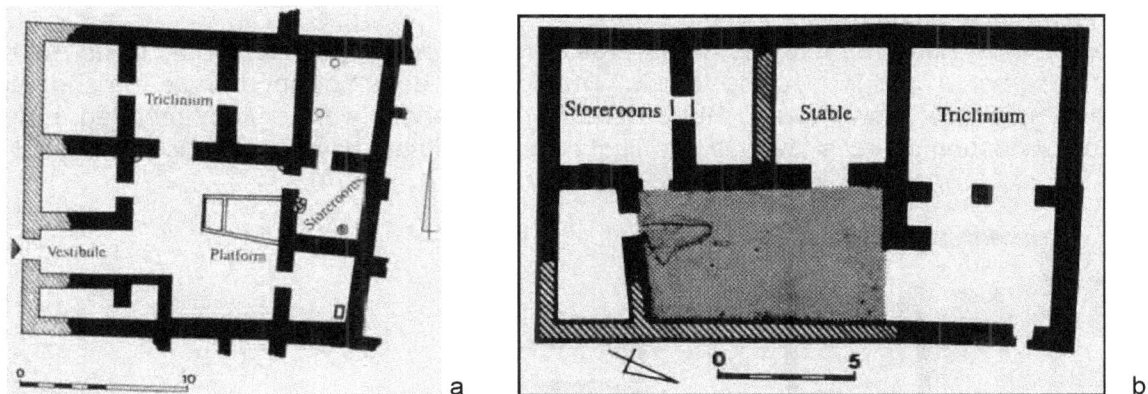

Figure 2. a. plan of the courtyard house under the Archaeological Museum, Amman Citadel (After Harding, 1951; Hirschfeld, 1995: 84, Fig. 60); b. plan of House 'G' at Pella
(Source: After McNicoll et al, 1982),

At least six courtyard structures dating to the seventh and eighth centuries were completely destroyed in the 749 earthquake. From a socio-economic aspect, generally, the houses at Pella represent the mixed-use function at ground floor level of the household: living arrangements accommodating animal stables storage of foods, workshop production, and some aspects of daily living (cooking, transit accommodation). In the upper floor spaces, much of the social activities take place and perhaps three houses at least with roof-top access. (Walmsley, 2007: 131). The upper floor could be reached through the courtyards by means of stone-built staircases (Walmsley, 2008: 251).

In one of the well-preserved examples of these houses, a two-storied courtyard, house 'G' (230 m^2) (Hirschfeld, 1995) (Fig. 2b), has a corner entrance leading to a simple rectangular courtyard to the east. The rooms on the lower level were also used as storerooms and stables. The presence of carbonized wooden beams suggests that the roofs were made of matting over oak beams sealed with clay (Walmsley, 2007: 130). The upper storey floors may have been carried on timber joists (McNicoll et al, 1982: 131).

An out of the ordinary house dating back to the late 7th century, destroyed by the severe earthquake of 749, represents a fine example of an urban, but not primarily residential complex. The complete ground plan remains unknown. It was a large complex (560 m^2) with two courtyards. The front façade of the house has three doorways opening directly onto the street (Fig. 3a). The group of living rooms in the west side of the house has accessibility from the main entrance through a small entrance hall. The eastern entrance was used to connect the two courtyards, while in the western side, a separate space was probably also used as a shop (McNicoll et al, 1982).

Socially, the excavators explain the parallel existence of the two courtyards due to the extended family's daily life activities that occupied the house, of which the closest courtyard to the street belongs to the men's wing. The large room built in the outer courtyard was a guest room while the inner courtyard and the rooms surrounding it might have served as the women's wing.

The "Umayyad House" in Jerash

An Umayyad residential quarter was found recently on the north side of the South Decumanus inhabited from 660 to 800 AD (Gawlikowski, 1986: 107-136). Socially, this also large Umayyad structure of about 600 m^2, coexists as 5-6 separate units, belonging to families that shared the same courtyard. The dwelling units laid around a courtyard are with one main entrance through a passageway from the colonnaded street in front of the house, which remains were in use, serving its original purpose along the lines of the shops.

The complex, however, extends northwards behind three shops that directly faced the street south of *decumanus* and formed the façade. The complex does not appropriate the shop space for its residential use (Gawlikowski, 1986: 111, 113). The shops were entirely restored, including the upper foundation courses found in the fill of a cistern without any major change in layout (Fig. 3 b and Fig. c).

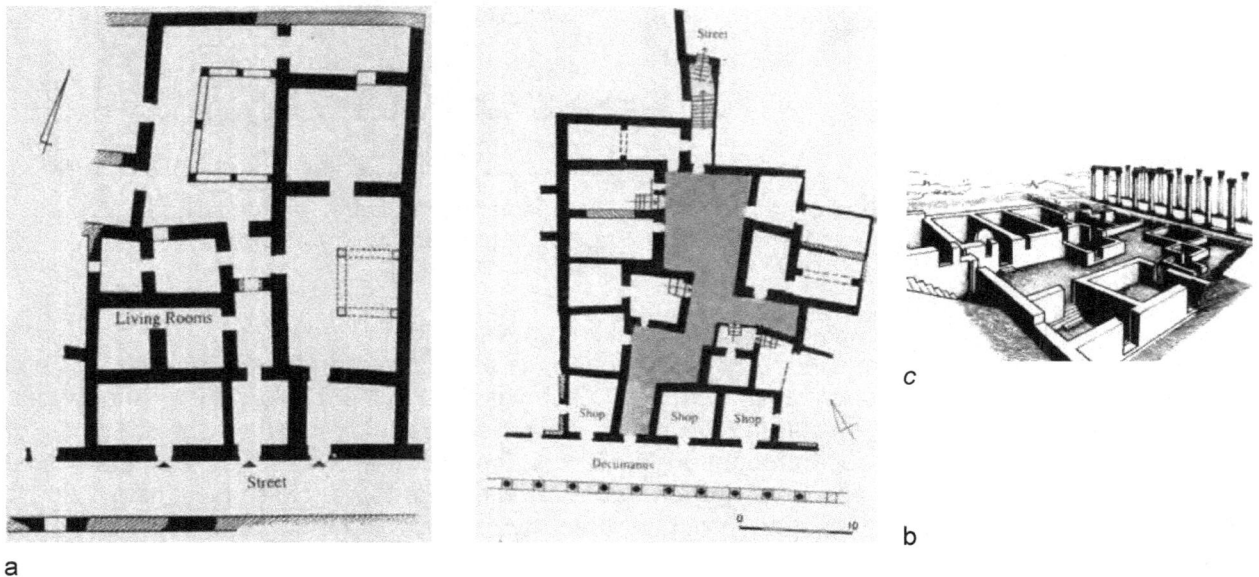

a

b

c

Figure 3 a. plan of the apartment house at Pella, seventh-eighth centuries (After McNicoll et al, 1992; Hirschfeld, 1995: 49, Fig: 25); b. plan of the Umayyad apartment house at Jerash (After Hirschfeld, 1995: 50, Fig. 26), c. 3D restitution views of existing state after restoration
(Source: After Gawlikowski, 1986: Fig. 2).

The entrance passage led directly from the street to an irregularly shaped courtyard. In the back of the courtyard, there was another opening that led through a staircase to the street north of the complex. The courtyard's irregular shape was the result of the intersection of the Roman period foundations' walls with the Umayyad period, as it is clear by the room that intrudes into the middle courtyard space (Gawlikowski, 1986: 113).

In this complex, there is no indication of the so-called bayt layout. The rooms are arranged into two wings, to the east and west of the courtyard, where the depths of the rooms of the west wing vary according to the pre-existing conditions that the builders encountered in the area. The eastern wing also is not arranged symmetrically and many rooms are not aligned on the same axis. The arrangement reflects the concept of the 'day and night' use of the living quarters as rooms are grouped in pairs; there are three sets of two-room suites (Gawlikowski, 1986: 114, 419). The front room earmarked for daily use and the back darker one used for sleeping. The layout of the units/ apartments, however, reflects a homogeneous pattern.

A sewage drain extends from the end of the courtyard to beneath the entrance (Gawlikowski, 1986: 113). This serves as the only sanitary facility in the household. An earlier sewage drain runs from the end of the courtyard and beneath the entrance. Some walls are preserved up to 3m above the floor, though the ceiling could not be lower than about 3.5m. The walls were probably mud-plastered while the roof is supported by wooden beams. An upper storey may have existed, but no evidence of it was found (Gawlikowski, 1986: 114).

In conclusion, comparing the main architectural features of the three previous houses reflecting the different socio-economic and political associations, we can note that the Amman complex example bears several significant differences and has only one main common feature with those of Jerash and Pella: the main façade entrance, which in this case is strongly connected to the courtyard through a vestibule. Also, according to Almagro (1992, 351), in each of these "Umayyad houses", several constants are apparent. He concluded that each has a courtyard, generally irregular in form, which functions as an element of distribution. All of the rooms to the house have either direct or indirect access to the courtyard, where at least one of the main rooms opens to the courtyard directly. Access from outside the house or from the street is gained through one sole exterior door and a series of hallways and small vestibules. However, in the Amman house case, the entrance from the outside to the courtyard is direct, since no L-shaped passages are used to obstruct the vision of the visitor. Another common feature is the hierarchal arrangement in the remaining rooms, of which many are only indirectly connected to the courtyard by way of other rooms.

However, the main difference with the Amman example is that there is a bayt layout at the main unit. According to Northedge (1992, 157), the bayt layout arrangement is directly paralleled by the proto-bayts at Khirbat al-Baydā, and the addition of a bayt in both households at Amman may be explained by the fact that these houses correspond to a part of the Umayyad political citadel project, which represents a single planned unit whose elements include the palace, the rebuilding of the fortification circuit, the open cistern, the Stratum V buildings of Areas B, C, and the Museum site. Whatever form and internal arrangements it may have had, it was in substantial use in the seventh and eighth centuries. At the same time, it reflects some socio-economic conditions and relationship with urban domestic architectural traditions of the late Antiquity era of Bilad al-Sham.

Another difference, probably in relation to the extended family's needs, can be seen from their size: at Amman is 380 m^2, at Tabaqat Fahl 560 m^2, and at Jerash about 600 m^2. The Tabaqat Fahl example, in fact, has many features in common with the one at Jerash: façade entrance between shops, no indication of the bayt layout, the ratio of the length of the façade to the length of the house is about 1:1.5, and the main architectural concept layout is the outcome of two units separated also by a forced earlier phase.

On the other hand, the "Umayyad house" in Jerash, given its irregularity, shows that the Umayyads dealt in a creative respectful approach to the potentiality of pre-existing features, achieving the basic religious, socio-cultural, and economic conditions and requirements. All of the windows face the courtyard suggest an inward orientation to maximize privacy, as also the design of the main entrance to a passageway, which turns at a right angle, obstructing direct view into the courtyard space from the street (Ahn, 2010: 106).

The main layout reflected in the architectural concept, in fact, is the result of two opposite approximately triangle-shaped units separated by an irregular forced courtyard. Comparing the Tabaqat Fahl house, which is not primarily residential, to the "Umayyad house" in Jerash given its irregularity, we can assume that there is a possibility that it might also have not been designed solely a residential house.

Basically, the then present socio-economic conditions played a major role in these urban sites, for there are no modifications in the region's urban living style during the Umayyad period given the eminent sense of religious tolerance inherent to the Islamic faith (Piccirillo, 1984).

THE AL-HALLABAT ARCHAEOLOGICAL COMPLEX AND THE AGRICULTURAL ENCLOSURE

The al-Hallabat archaeological site (Fig. 4) within the complex of the Qasr is located 60km northeast of Amman (Arce, 2007: 325), 25km to the northeast of the city of al-Zarqa on the southeast edge of the modern town of al-Hallabat al-Gharbiyya (Ghrayib, 2003: 65), and about 16km from the Via Nova Traiana (Kennedy, 2000: 90). Al-Hallabat was built on a gently sloping ground dissected by shallow rainwater gullies that drain the land to the south. The site lies on the top of a mound situated in a semi-arid zone with an annual precipitation rate of less than 100mm (Bisheh, 1985: 265).

This unique site was a Roman fortress with a *probable* Nabataean predecessor converted into a desert palace, and was rebuilt several times as attested by several identified phases of development (Kennedy, 2014; Kennedy and Riley: 1990; Bisheh, 1985; Arce, 2007). More analytically, the Qasr history goes back to the Nabataean period when it was a station on the trade routes. During the Roman period, it was a Roman fort constructed in the second or third century AD, as a military station on the road between Bosra and Aqaba (Harding, 1984). Built from black basalt and honey-colored limestone (Kennedy, 2000: 90), it dominated the site to monitor and control a broad area to the southeast towards 'Azraq from which travellers would be observable for many kilometers while approaching the plateau along the Amman-Busra-Damascus route (Ghrayib, 2003; Jalboosh, 2009).

An Umayyad mosque also dominates the site from the top of the mound and several Umayyad houses remains are still visible on the slopes of the mound and in the valley. The impressive architecture of the Qasr, the mosque and houses, which belong to the same period are unique examples of Umayyad rural Jordan.

The archaeological site covers an area of 50 acres ($202342.821m^2$) (Ghrayib, 2003: 65; Jalboosh, 2009). However, according to Kennedy (2014: 107, Table 1) the area of the 26 structures is about c.35 hectares ($350000m^2$) and with dimensions c.850m x 550m= $467500m^2$. An agricultural enclosure is located about 400m to the west of the Qasr. The enclosure of about 270m x 220m = $59400m^2$ collected the water that reached it from two wadis (Bisheh, 1982: 142). It is irregular and gradually narrows to the lowest point of the ground elevation on the north, the walls of which only one course of stones remains, were built of rubble core of field stones without a foundation trench (Bisheh, 1982: 138).

This Umayyad 'Nomad Village', according to Kennedy (2014: 108), seems to have been placed in a good region for cultivation and probably remained largely based on animal herding. Excavations of a number of sluices and water deflectors, however, confirmed that this was an agricultural settlement (Bisheh, 1980: 70).

The agricultural enclosure associated with the site has an elaborate system of sluices regulating the distribution of water to its plots (Bisheh, 1985: 264-265). It is described by the present inhabitants of al-Hallabat as 'Huwaytah' (diminutive of Hait) (Creswell, 1989). 'Hait' is a word used in medieval texts to denote cultivated areas or gardens around a town (Grabar et al, 1978). In addition, the evidence of the stone and the basalt objects used for grinding and processing seeds and vegetables attests that the inhabitants of the area were depending on agriculture. The existence of two stones in one of the ruined buildings, to the west of the water reservoir, also suggests that the enclosure was devoted to the cultivation of orchards, containing mainly olive trees and vines. Actually, agricultural improvements instigated by the Umayyads resulted in the spread of agricultural settlements (Ahn, 2010: 102).

The site is also located in an area of numerous springs and water sources and includes a complex water system with channels; at least five large cisterns and a big reservoir cut in the bedrock down in the valley (Arce, 2007: 325) and an elaborate bath complex (Hammam as-Sarah) display the Umayyad celebration of their water infrastructure and their control over water resources (Alhasanat et al, 2012: 357).

The reservoir (2060m^2, volume of 8000m^3) lies a few hundred meters to the south and the numerous cisterns in the wadi to the north and west and the channels system were probably connected in order to store the water and distribute it to the Qasr, the houses, and the agricultural land (Ghrayib, 2003: 68; Bisheh, 1989: 246; Harding, 1984). These various structures seem to be randomly scattered around the Qasr (44x44m), for the most part facing south and /or east, while the houses were built along very ordinary, seasonally-flooded wadis in an essentially featureless landscape.

The particularity of al-Hallabat settlement is that it has a pre-existing Qasr, located on the top of the mound, and later on was surrounded by the houses and hydraulic system. Looking to other Umayyad settlements in the region, mostly the Umayyad palaces were built on a flat area, without houses surrounding them, such as Qasr Kharana and Mushatta, or we can find a small flat settlement without a palace such as at ar-Risha.

On the other hand, the unexcavated Khirbet al-Askar (c. 33km south-east of Karak and 10km east of Muhai), according to Kennedy (2014: 107, Table 1), has the same area (c.35ha) with dimensions (c. 1100mx350m=385000m^2), but with 45 structures.

An analogous situation, until now, to our case is found at the al-Qastal south of Amman, and Jabal Seys in Syria where the Qasr is surrounded by houses and other installations.

As noted by Kennedy (2014: 99), contradictory to ar-Risha ruins (with an area of c. 5ha, and with dimensions of c. 400x180 m=72000 m^2 (Kennedy, 2014: 107, Table 1) (Fig. 9c), where "some buildings are arranged in short lines and are roughly parallel" and fairly compact, while at al-Hallabat (Fig.4) as also at Jabal Seys,Qasr as-Swab, and Khirbet al-Askar, there "is no order in the layout, and structures are seldom aligned, even with a neighbouring building, and they are widely dispersed" across the site, and in all these three cases shape "an elongated settlement — long and narrow and covering a considerable area" (Kennedy, 2014: 107).

This phenomenon of covering a considerable area at both sites of al-Hallabat and Khirbet al-Askar might be explained due to the similarities of the socio-economic existing conditions at that time. Small farms and garden areas between the houses were the main characteristic feature of the land use formation, similar to that of the farmhouse at Nahal Mitnan (Haiman, 1995, Plan 3) (Fig. 8c). In Nahal Mitnan, a 4km long wadi channel was intensively cultivated by means of agricultural terraces and about two km upstream from its confluence with Nahal Horsha (Haiman, 1995: a:1). The Nahal Mitnan farm consists of the main farmhouse, an agricultural installation, a threshing floor, and a section of the terraced wadi-channel, enclosed by a stone fence (Ahn, 2010: Fig. 44 and Fig. 45).

This might also suggest that we should accept that the two sites of al-Hallabat and Khirbet al-Askar are more suited and specialized for large cultivation centres, but are also more of organized centres compared to ar-Risha (Kennedy, 2014: 101), even as clarified before that al-Hallabat site is a far more developed site than ar-Risha.

Socio-economically, this might also suggest that Bisheh's (1985: 264) identification, that probably those houses were the residences of servants working in the Qasr, was overestimated. One cannot accept that such complex houses, with considerable sizes, such as the well-planned house no. 1 (724m^2) of al-Hallabat, which is far away about 400m of the Qasr, belonged to servants. Furthermore, this house is directly connected to an agricultural enclosure (see fig. 4).

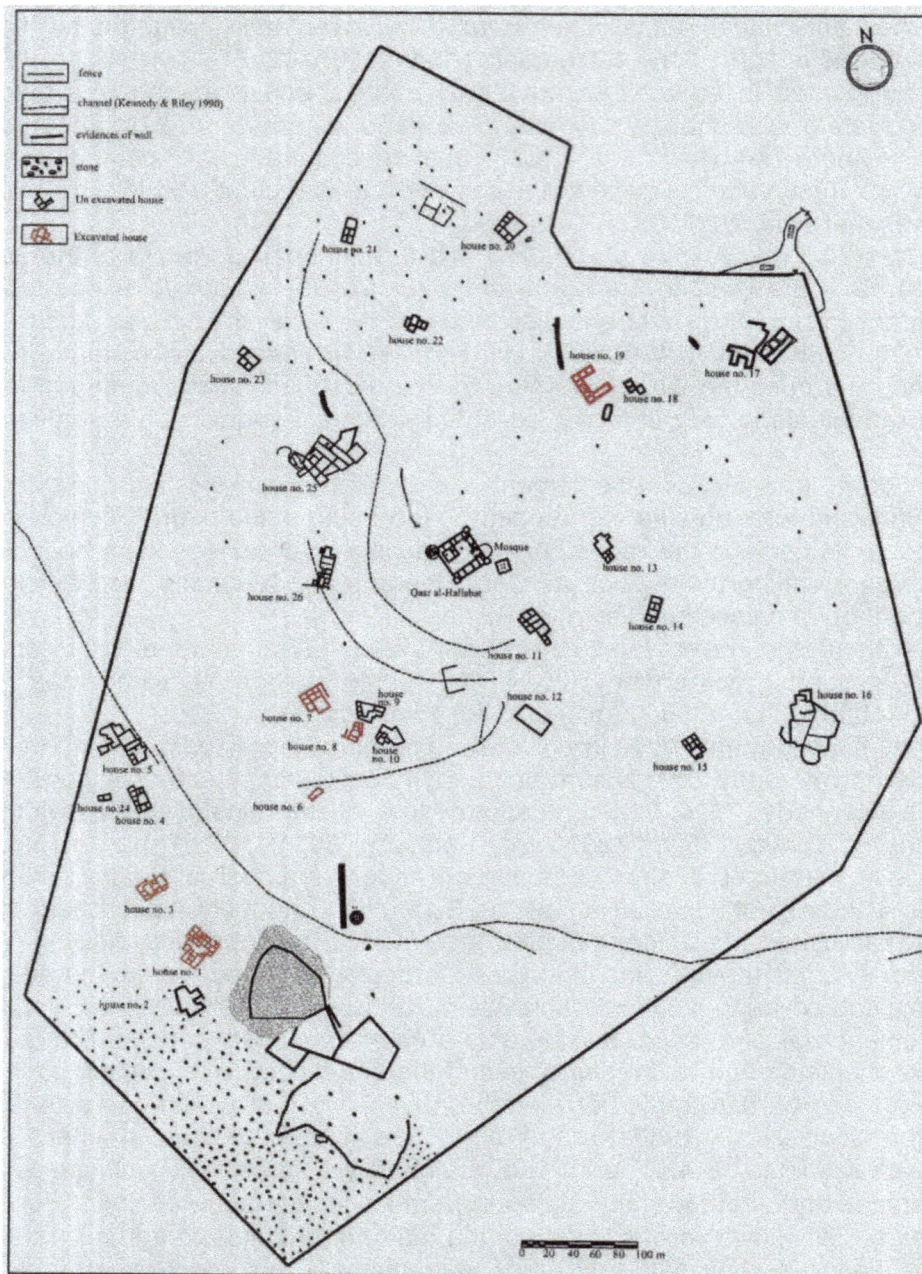

Figure 4. Site plan of the Umayyad houses around the Qasr at al-Hallabat (Source: Authors).

THE UMAYYAD HOUSES AT AL-HALLABAT SETTLEMENT ARCHAEOLOGICAL SITE: SOCIO-ECONOMIC CONDITIONS

The twenty six Umayyad houses/ structures (Table 1, Fig. 5), of which six houses have been recently excavated and the other twenty surveyed, can provide valuable data for establishing some of the socio-economic conditions and needs based on their typological patterns and classifications in relation to the various Umayyad Jordan house-types. The study of pottery confirmed that these belong to the Umayyad period and were built in a limited period of time as proven by the type of vessels, which were reserved for domestic purposes, like cooking pots, jars, bowls, casseroles, and storage jars (Ghrayib, 2003: 67; Jalboosh, 2009). The following presented data was organized based on unpublished and published reports of these houses, in addition to preliminary excavation results. The twenty six houses have been identified and numbered (Table 1).

In all of these houses built according to the site topography, we can find many similar architectural features. The majority consist of a group of rooms surrounding the open courtyard (Fig. 5) with a well-planned water distribution system, which served the entire settlement. However, many houses had been transformed through the times, since evidence of enlargements has been observed (Ghrayib, 2003; Jalboosh, 2009).

Although they are better built than at ar-Risha (Kennedy, 2014: 101), the layout is almost little randomly scattered on the slopes around the palace and beside the large reservoir, and the well-planned water distribution system served the entire settlement by a thorough network of channels. However, every house had a cistern or a well nearby. Bell-shaped cisterns had been dug into the bedrock and were completely plastered. Meanwhile, the water supply was irregular during the rainy season and where the flow of water was abundant, several protective structures, such as wells and an earthbound, were built around the reservoir to collect the surplus water (Fig. 4).

Table 1: Classification and typological patterns of houses at al-Hallabat settlement.

House No	Area (m^2)	State	Type	Room No	Courtyard No	Function
1	724	excavated	Complex	24	3	Residential
2	517	Un excavated	Complex	23	3	?
3	362	excavated	Complex	11	2	Residential
4	280	Un excavated	Courtyard	9	1	?
5	721	Un excavated	Complex	11	3	?
6	72	excavated	Nucleus	1	1	Storage
7	290	excavated	Nucleus	7	1	Workshops
8	229	excavated	Nucleus	6	2	Residential
9	322	Un excavated	Courtyard?	?	1	?
10	375	Un excavated	Nucleus	?	1	?
11	430	Un excavated	?	9	?	?
12	440	Un excavated	?	?	?	?
13	279	Un excavated	Nucleus	?	?	?
14	222	Ur excavated	Nucleus	4	2	?
15	269	Ur excavated	Nucleus	8	1	?
16	2200	Ur excavated	Complex	25	2	?
17	1435	Ur excavated	Complex	8	1	?
18	190	Ur excavated	Nucleus	3	1	?
19	633	excavated	Courtyard	8	1	Khan
20	420	Un excavated	Nucleus	6	1	?
21	220	Un excavated	Nucleus	4	1	?
22	190	Un excavated	?	4	2	?
23	230	Un excavated	Nucleus	3	1	?
24	61	Un excavated	Nucleus	1	1	?
25	1558	Un excavated	Complex	27	4	?
26	465	Un excavated	Complex	24	3	?

Al-Hallabat houses were built without foundation trenches, directly on bedrock and gravel surfaces that sloped gently. The building material is a stone of different kinds, mainly limestone and re-used basalt blocks and fieldstone. The most common floor type is compact earth in both rooms and courtyards. Most of the houses had flat roofs provided with drain pipes and drainage channels dug parallel to the walls of the houses, but archaeological finds suggest also the use of tiles in few houses.

Staircases, actually, to the upper stories were common in the Umayyad period. They formed a vital element in the ordinary house tradition of Hauran. No constructed stone staircase was found, but wooden stairs and ladders may have given access from the courtyards to the roofs, and even to the second storey of living rooms, as will be discussed.

No signs of directing surface runoff were also observed on the ground, although certain structural expediencies in some of the buildings' details suggest that the foundations may have been protected from erosion. All structures were built on about the same absolute level. No building stood any higher, or on remarkably better grounds than another except building no. (19), built in front of the Qasr façade. However, this house has a totally different scale and typology, as shall be shown.

According to Ghrayib (2003), the houses can be divided into two types: residential complex and simple houses. Their typology, however, is featured by two main schemes reflected in their architectural layout; complex houses and isolated houses. The houses, however, reveal primary differentiation between dwellings, ranging from relatively simple one-room structures to complex multi-family dwellings. The rooms include working areas with *tabuns* and other installations, which might have been small open courtyards. The rooms are aligned along an open space that may have provided access to more than one familial unit.

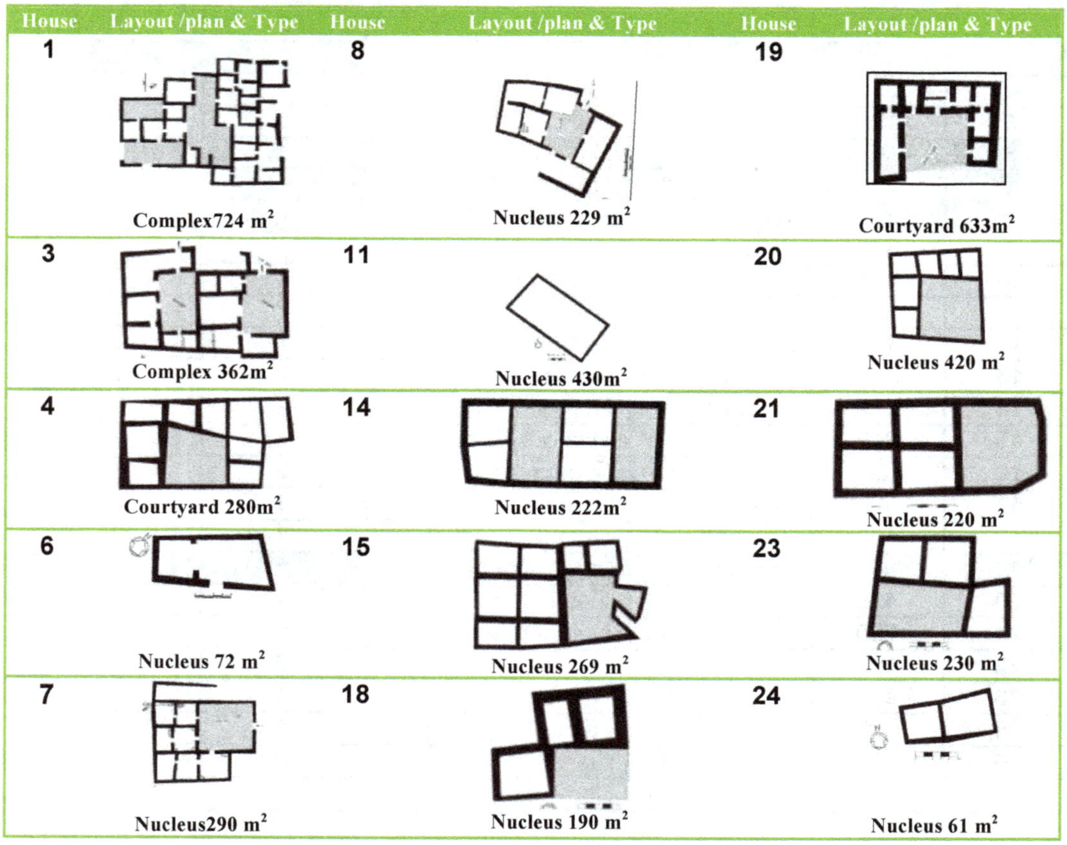

House	Layout /plan & Type	House	Layout /plan & Type	House	Layout /plan & Type
1	Complex724 m²	8	Nucleus 229 m²	19	Courtyard 633m²
3	Complex 362m²	11	Nucleus 430m²	20	Nucleus 420 m²
4	Courtyard 280m²	14	Nucleus 222m²	21	Nucleus 220 m²
6	Nucleus 72 m²	15	Nucleus 269 m²	23	Nucleus 230 m²
7	Nucleus290 m²	18	Nucleus 190 m²	24	Nucleus 61 m²

Figure 5. General plans of most of the Umayyad houses around the Qasr at al-Hallabat (Source: Authors).

DISCUSSION AND RESULTS: TYPOLOGICAL STUDY OF AL-HALLABAT HOUSES IN RELATION TO SOCIO-ECONOMIC INDICATORS

According to Petruccioli (2007), the meaning of a rural house can be defined as an expression of a vital, useful realistic architecture that responds to everyday, practical needs, such as shelter, warmth, storage of food, in addition to protection of domesticated animals. He (2007: 68) states that "it avoids gratuitous innovations, uses the simplest techniques to ensure a certain stability, and efficiently meets the most basic of family needs". From a socio-economic aspect, Petruccioli (2007: 68) assumes that, in the early Islamic period, there were many pieces of evidence for a process in which urban houses become ruralized. However, according to Ahn (2010: 101), "the relationship between dwelling and place is first established in a rural setting, though, the first urban building systems are influenced by their rural counterpart".

However, with types of buildings already known in a rural context progressively introduced into the city, Polci (2003: 101) also argued that the rural areas' vitality may have contributed to urban space becoming ruralized. For example, some house unit layouts of rural settings especially in Umm el-Jimal, Subaytah and Msayké features many common elements such as the flanked by rooms central private courtyard used for a variety of functions, with upstairs living quarters reached only by a courtyard staircase (Walmsley 2007:132). Many of al-Hallabat rural settlement complex houses can also sustain this suggested process model as will be shown. Actually, the socio-economic conditions of the Umayyad rural settlement expansion, as can be seen from the al-Hallabat agricultural settlement, may have also created an atmosphere conducive to a process of ruralization of urban space.

On the other hand, the interior courtyard house is an expression of notions of privacy dictated by religious and social norms (Ahn, 2010: 107). According to Hakim (1986: 95-96), the courtyard house creates a physical setting suitable for the religious and social requirements of Islam: privacy, interdependence, and Bātin vs. Zāhir. However, both natural and cultural factors affected directly or indirectly the design layout of these houses. So this section will summarize our understanding of these houses' basic design components: entrance, interior arrangement, and courtyard location.

Based on the architectural layout, the discussed houses at Amman, Jerash, and Pella's urban centers, as also the mentioned early Islamic "Nomad Village" settlement at ar-Risha, in which fifteen buildings extended for some 300m (Kennedy 2014: 99) and al-Hallabat complex archaeological site shall be discussed in the following section. These can be broadly divided into two main categories: the complex house and the courtyard house.

The complex house can be divided into two sub-groups: a) the urban complex house, such as the houses at Pella and Jerash, created by the construction of adjoining dwelling units around a common courtyard with shops, and b) the rural farmhouse, consisting of several dwelling units and wings composed around at least one courtyard. However, these types of houses were mainly found at ar-Risha, Khirbet al-Askar, and at al-Hallabat settlement, as will be illustrated later on. For the complex house of both sub-groups, it is difficult to determine whether the compound was formed gradually or was originally planned as a single complex.

For the courtyard house type, such as in the citadel of Amman, the central courtyard is without pillars, a feature that characterized the domestic urban buildings of Umayyad Jordan. This type provides links to the early Byzantine architectural traditions of the region where the local courtyard type continued up to relatively recent times. However, in Jordan, both the Umayyad courtyard and complex house have several common features. Each has a courtyard which functions as an element of distribution. While all of the house's rooms have either direct or indirect access to the courtyard, at least one main room opens directly to the courtyard.

Accessibility to al-Hallabat houses was usually through a single entrance. This entrance was a vital link between the courtyard and the other spaces of the house, as also a source of light and air. Because of the need to leave the house entrance open sometimes during the day (for light and ventilation as also for the nature of agricultural works), the entrance to the courtyard was usually positioned at a different axis from that of the main entrance to the units. This arrangement prevented a glimpse by passers-by into the private areas of the house interior. So, in light of the prominence of the house entrance, builders usually took great care in constructing the doorframe. The threshold, doorposts, and lintel formed a sturdy structural unit intended to enhance not only the stability of the house as a whole but also the socio-economic class of the household.

The mixed function of households is also evident in al-Hallabat houses, where living arrangements could also accommodate animals. In particular, the significant evidence for troughs and managers in both urban and rural houses suggest the prevalence of animal stabling in the early Islamic period, as in the main house over the Museum site at Amman, where rectangular and circular limestone troughs were found (Harding, 1951: 9). However, there is little evidence of manufacturing activities taking place in al-Hallabat structures in relation to the evidence of agricultural activities.

In addition, the sewage drains, open drains, and the cisterns of the al-Hallabat houses are concentrated mainly in the area of the courtyard, as can be seen in the urban houses of Amman and Jerash. The courtyard space in the internal organization of most of these houses is corroborated by the layout of the rooms around the courtyard; most of the rooms open directly onto the courtyard. This arrangement is exemplified also by household 'G' at Pella, which is organized around an internal courtyard with five doorways around the courtyard giving access to surrounding rooms (Fig.2 b). The courtyard as the nucleus of the house also dictates that the main entrance leads from the outside to the courtyard.

Table 2. Rooms size classification at al-Hallabat settlement.

House No.	Rooms No.	Large rooms No. (7-10 m length)	Medium rooms No. (5-7 m length)	Small rooms No. (3-5 m length)
Complex no.1	24	4	10	10
Complex no.3	11	2	4	5
House no.6	1	0	1	0
House no.7	7	7	0	0
House no.8	6	1	3	2
Structure no.19	8	8	0	0
Total	57	22	18	17

Generally, the rooms' width from the six excavated al-Hallabat houses range from 2.5-4.5m due to the limited availability of wooden beams for roofing, meanwhile their length range from 3-10m (see Table 2). On average, the rooms are relatively of medium size (3x4m). This is a characteristic element of the Umayyad houses at al-Hallabat. Based on Table 1, which presents the 26 structures' classification, their layout and their courtyard location (Fig. 4 and Fig. 5), we can categorize them into three main types that might also reflect the socio-economic conditions of the settlement householders: nucleus, complex, and courtyard type. The following are their typological features' and characteristics:

The nucleus type (60m^2 - 430 m^2): The common house

Twelve nucleus structures out of the 26 were identified. These are no. (6, 7, 8, 10, 13, 14, 15, 18, 20, 21, 23, 24) (Table 1). However, out of the 26 structures, 3 were not classified. These are no. (11, 12, 22). The five structures (7, 15, 18, 20, and 23) (Figs. 5, 7a, 8a, 8b, 9a) are similar in their (L) shape layout and lies on a different alignment. It is noted that the two buildings (14 and 21) are characterized by the same layout, based on the room numbers (4 rooms), location, and dimensions (220m^2). Another two are with irregular shapes (8 and 10), each with a different layout built near to each other. Structure no. (8) (12.68mx18.05m= 229m^2) consists of 6 rooms and 2 courtyards (Fig. 8a), located on the south slope mound where the Qasr is located. This building was used for habitation. However, it seems that this house was used in two phases. This, while the west side was built better than the east, in which a new door was opened at the north wall and later was closed, in addition, the east side is higher than the west. Two marble columns were discovered inside the house and could have been brought from the Qasr (Jalboosh, 2009; Ghrayib, 2003). There are also two rectangular stone platforms divided into two squares by a wall in the middle of the two other nucleus structures (6 and 24), with the same approximate dimensions, now preserved as heaps of rubble, but their original function is unclear.

This type is the most basic and is commonly used by the vast majority of the settlement inhabitants. These are generally isolated houses. The basic simple type consists of one-roomed structure or side-by-side units, built either behind or in front of an open courtyard, thus the courtyard is alongside the house.

The floors are of beaten earth and the walls are constructed of stone on the bedrock. The courtyard adjacent to the house was generally spacious and open to light and air. Sometimes, especially in houses built on a slope, the courtyards were placed together between the closely built houses.

From a socio-economic point of view, its advantages are obvious, since the courtyard both created a convenient barrier between the public and private domains and provided an additional out-of-doors working area, usable for much of the year in the generally high temperate climate of the site. A variant of the simple type is the two-wing house. This sub-type has two perpendicular wings, usually built in the northern and western part of the courtyard.

The complex type (360m^2 - 2200m^2): The extended family house

Eight buildings, each with a different layout no. (1, 2, 3, 5, 16, 17, 25, and 26) were identified as complex house types (Fig. 5, 10a, 11a). They can be divided into two groups: a) the so-called "urban apartment house", including several units around a common courtyard, or around several (2-4) courtyards, and b) the estate house, including several units and various wings arranged around a spacious central courtyard. In both cases, the original houses were clearly enlarged to suit the needs of the extended family. They have more than one courtyard and used by multi families and built on the site edges. The main features of these houses can be found in house complex no. (1) (724m^2) (Fig. 10a) located about 30m to the north-west side of the reservoir, and the rectangular complex no. (3) (362m^2) (Fig. 11a) located about 80m west of the complex no. (1), which contains two residential units west and east, in which the east one is built much better than the west. This type is an expansion of the simple nucleus house. It includes several units and various wings. It consists of rows of single rooms and sometimes a pair of rooms, one behind the other as in the urban houses. The units are clustered around three or more sides of series of courtyards, or common open spaces. The units are aligned according to the topography, running roughly north-south along the slope of the mound.

The courtyard type (250m^2 - 400m^2): The public building

Three courtyard structures were identified (4, 9, and 19) (Fig. 5, 12a, 12b). The structures (4 and 9) have different layouts, but with courtyards surrounded by series of rooms from three sides. The building no. (19) (633m^2) (Fig. 12b), located in front of the main façade of the Qasr is also characterized by a (U) shape courtyard, but based on its large size compared with the other houses, we can assume that it is of a special type and had a different function than the others; it could have been used for a public activity of a khan, as will be shown.

COMPREHENSIVE TYPOLOGICAL PATTERNS' CLASSIFICATION ANALYSIS OF AL-HALLABAT HOUSES WITH BILAD AL-SHAM EXAMPLES

The purpose of this section is to identify, based on the comparative study of al-Hallabat houses and other houses, whether there were common typological patterns in the region of Bilad al-Sham, according to their layout, size, and room arrangement around the courtyard. This analysis will also provide a more holistic picture of the socio-economic conditions based on the three housing types at al-Hallabat.

Nucleus House

It should be noted that this type was not always small in scale and in the architectural sense while many are impressive in their size and quality of construction. Such houses measured between 61m^2 (house no. 24) and 375m^2 (house no. 10). Characteristic examples are houses no. (6, 21, 20, 18, and 23).

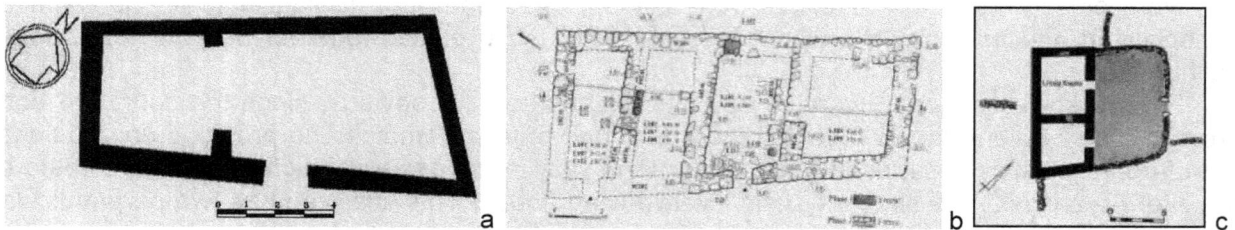

Figure 6. (a) plan of house no. (6) at al-Hallabat,(b) plan of building at Ein 'Aneva (After Magness, 2004) (c) plan of a simple farmhouse at Nahal ha-'Etz (Source: After Hirschfeld, 1995: 38, Fig.14).

House no. (6) is a rectangular building (12x6 m) with one room and threshold door made from compacted clay (Fig. 6a). The walls (0.76m) are made from large and medium field lime- stones. Analogous exterior shape and size with this house was found in a small rectangular house (6x15.5m) built near Ein 'Aneva (Fig. 6b) at Nahal Zeelim (Wadi Seiyal) about 4km north of Masada. However, it is with two habitation units aligned on the same axis oriented north-south. The floors were made of packed earth, laid over a fill (Magness, 2004).

Interesting enough, there were earlier similar but simple examples of a farmhouse, dated back to the Byzantine period (5th - 6th centuries), found at Nahal ha-'Etz in southern Palestine. The structure is rectangular (67m^2) with two wings used as living rooms (Fig. 7c). Its walls and fence surrounding the courtyard were integrated with the agricultural terraces preserved in the streambed.

Such architectural units often became the core of later enlarged farmhouses of the complex type (Hirschfeld, 1995). An analogous example of house no. (21) (Fig. 7a) was found at Horvat Susiya, southeast of Hebron, dated to the 6th century, and continued in use until the end of the 7th century (Fig. 7b). The structure (160m^2) had one storey with a courtyard (244m^2). It is divided into two symmetrical wings: a northern wing with three living rooms, and a southern wing with one large spacious room. The latter most likely served as the triclinium for family meals and entertainment. Two small shops were built at the back, facing the alley west of the house (Hirschfeld, 1995: 36).

From a socio-economic perspective, the Horvat Susiya and the structure no. (21) at al-Hallabat might suggest that these rural structures were provided with shops for direct goods and trade, as in the case of the Urban houses at both of the Urban centres at Jerash and Pella.

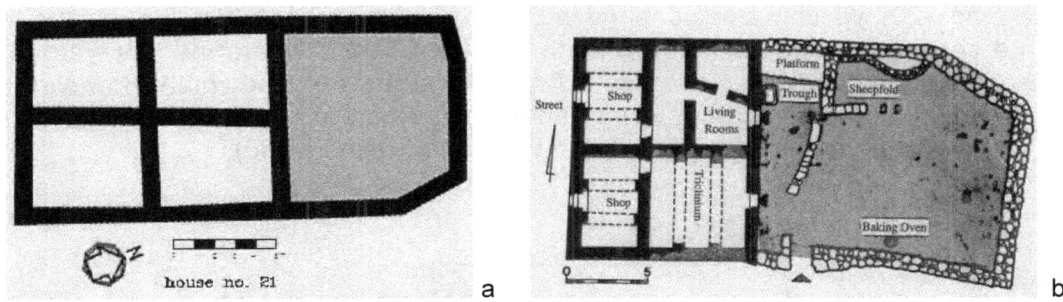

Figure 7. (a) plan of House no.(21) at al-Hallabat, (b) plan of the house at Horvat Susiya
(Source: After Hirschfeld, 1995: 3, Fig. 13).

Similar room arrangements in houses no. (18 and 23) (Fig. 8a and 8b) can be seen in the farmhouse at Naḥal Mitnan (Fig. 8c). Archaeological evidence suggests that the establishment and occupation of the farms was a process that continued through the sixth, seventh, and eighth centuries (Magness, 2003: 137). The agricultural system includes an extensive network of terraced wadis, numerous farmhouses, and various agricultural installations (Ahn, 2010: 41).

This farmhouse complex (15x33 m) is composed of three dwelling units, each consisting of one to three rooms and a small courtyard (Haiman, 1995a: 3). However, one can note that there is even less of a physical division between the living quarters and the area for agricultural and stabling activities of the household (Ahn, 2010: 90). The outer courtyards may have been utilized for activities related to the collection of crops before storage or transport, as well as for animals. The two main rooms containing raised beaten-earth platforms, used as beds, functioned as the sleeping quarters. Nevertheless they constitute one structure farmhouse presumably inhabited by three nucleus families (Haiman, 1995a: 3-4).

Analogous earlier but in a more organized structured layout was found also in the Nahal ha-Ro'a (Fig. 8d), a two-winged farmhouse (270m^2) dated to the Byzantine period (5th-7th centuries) including six rooms surrounding the courtyard from two sides. According to Haiman (1995b: 45), and as we can assume the same for al-Hallabat settlement, during the Umayyad period, the agricultural settlement in the Negev desert was motivated by two reasons: imperial policy that aimed towards protecting the frontiers by encouraging the establishment of agricultural settlements, and state-sponsored settlement of semi-nomads.

Finally, an analogous structure to the house no. (20) (Fig. 9a) seems to be in large use at Khirbet al-Askar, as can be identified from the many structures, such as KS 2, KS 9, KS 10, and KS 38 and others (Kennedy 2014:104-05, Fig. 8,10).

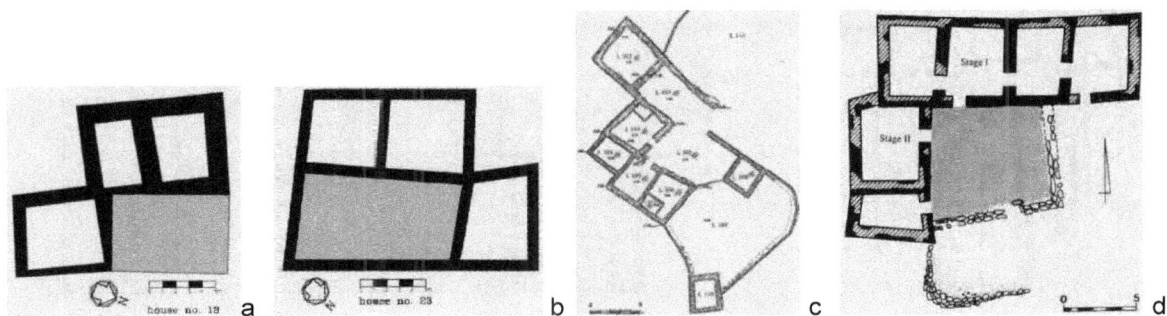

Figure 8. (a) plan of house no. (18) at al-Hallabat,(b) plan of house no. (23) at al-Hallabat, (c) plan of Nahal Mitnan the farmhouse (After Haiman, 1995, plan 3), (d) plan of farmhouse at Nahal ha-Ro'a
(Source: After Hirschfeld, 1995: 39, Fig. 15).

Similar to no. (20), which was also found in Structure J at ar-Risha (Fig. 9 b, c), is a two-wing house with a series of rooms built on the north and west flanks, where the courtyard (16x18m) was enclosed by a mud brick wall (average 0.7 m wide) at the east and south. However, no clear entry was visible, but since the east wall is unbroken, almost up to the north-east corner, the doorways into the courtyard must be located on the south (Helms, 1990).

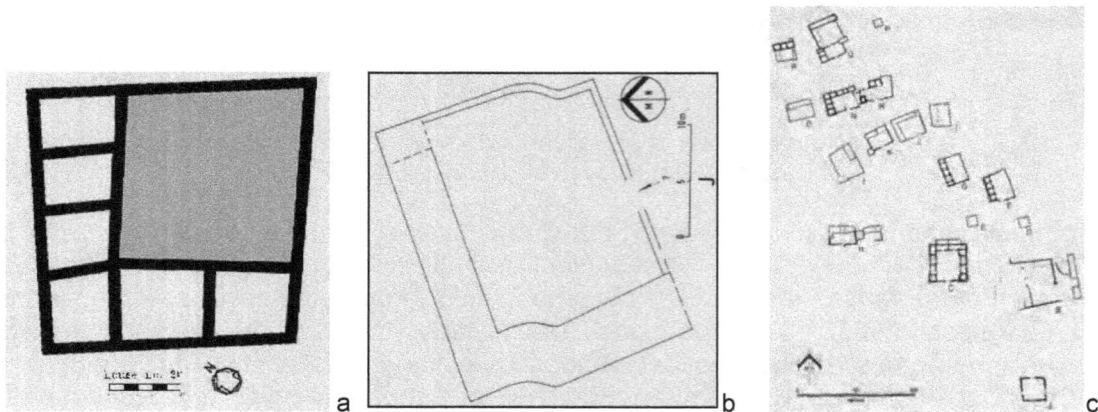

Figure 9. (a) plan of house no.(20) at al-Hallabat, (b) restoration plan of structure J at ar-Risha (After Helms, 1990: 115, Fig. 55), (c) site plan of structures at ar-Risha
(Source: After Helms, 1990: 129, Fig. 71 and 129).

Complex House
This house-type, from a socio-economic point of view, whether built to accommodate the members of a growing extended family or built as a result of the enlargement of the owner's property, offered greater privacy and protection for the activities conducted by the inhabitants, mainly in the courtyard. It also offered direct access from the courtyard to the adjoining public area. A characteristic example of this type is complex no. (1) with 24 rooms and three courtyards (Fig.10a), where the external walls (0.6-0.7 m) are made from large fieldstones well set and filled with smaller stones. This house (724m^2) has two residential units west and east. An analogous layout to this complex was found at the domestic houses XII and XIII at Umm el-Jimal (Fig.10b), located almost in the middle of the town, east of the "Cathedral", and west of the Roman reservoir which is also the case of al-Hallabat complex no. (1).

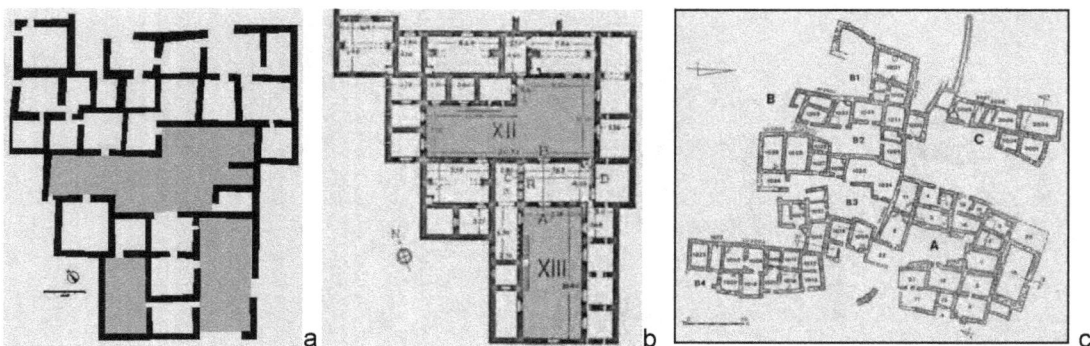

Figure 10. (a) plan of complex no.(1) at al-Hallabat, (b) plans of houses XII and XIII at Umm el-Jimal (after Butler, 1919: 203, Fig. III.182), (c) plan of Khirbet Abu Suwwana (Source: Magness, 2004).

In general, at Umm el-Jimal houses, much of the ground floor space is devoted to the agricultural activity, animals, stables, shops, and domestic activity, as indicated by the presence of the kitchen. The ground floor often contains wells, cisterns, and in some cases, a room provided with a basic drainage system. Living quarters are situated on the upper floor (De Vries, 1981: 63). According to De Vries (1998: 109), this arrangement of domestic space is not necessarily motivated by the separation of people from animals. The building layouts indicate an intimate sharing of space, with people and animals in constant contact with one another. This seems to be the case of al-Hallabat resident complex no. (1). The same can be said for the courtyard houses at Pella, which resemble the domestic units found at Umm el-Jimal.

On the other hand, to achieve the needed privacy at Umm el-Jimal and Pella, the stairs leading to the upper living quarters were located in the courtyards. This emphasizes the courtyard as a private rather than public space. The ground floor area of the houses at Pella (Walmsley, 2007: 131), as at al-Hallabat, focuses on the central internal courtyard, where much of the interactive daily life of the household takes place (like cooking and care for domesticated animals). This may also lead to an assumption that many houses at al-Hallabat had a second floor, such as in complex no. (1).

Actually, there are many other analogous structures to complex no. (1), such as what was found at Khirbet Abu Suwwana (Fig. 10c), an early Islamic village located near Ma'aleh Adumim, east of Jerusalem. The excavation revealed two types of residential units; one of them has a crowded system of residential units and a mosque, and the second has six residential units of high quality. The mosque could have accommodated up to fifty-four worshipers (Magness, 2004). Somehow, we can also observe similar applications to these houses and at al-Hallabat, such as walls covered sometimes with plaster (house no. 1 and 19). They were also built directly on the bedrock along the topography and most of the floors were of packed dirt and lime. Essentially, the excavations of complex no. (1) is not yet completed, and if further excavation reveals more walls extended from complex no. (1) to complex no. (3) (Fig.11a), then we might have analogous structure layouts such as those found in Khirbet Abu Suwwana.

Rather like complex no. (3), similar conditions can also be observed at the structure N/M located at ar-Risha. It consists of two parts: the western wing (N), and eastern wing (M) connected by the courtyard walls composing one complex (Fig. 11b). The western wing consists of two adjoining cell blocks, but it is more regularly set out, containing four cells each about 4x4m, in which the northern section is with four cells of irregular dimensions (3x3m - 5x5m) built against the western one. Wing M is integrated with wing N, but it is much simpler in the layout. The single cell block (4x4m) was accessed through a narrow door (0.75m wide) from a roofed vestibule (4x5m), which was apparently opening towards the courtyard (Helms, 1990).

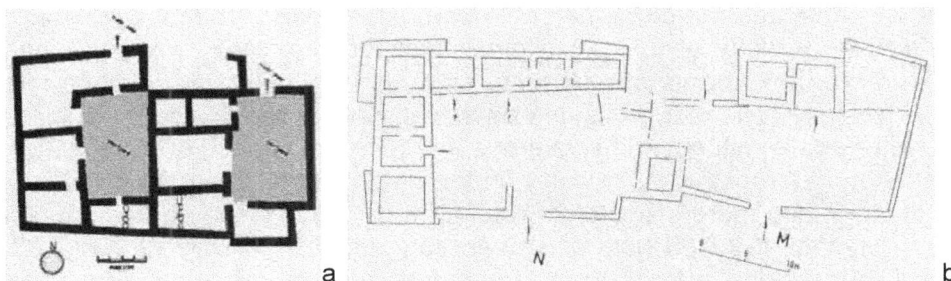

Figure 11. (a) plan of complex no. (3) at al-Hallabat, (b): Reconstruction plan of Structure N & M at ar-Risha (Source: Helms, 1990: 120, Fig. 61).

Courtyard House

This type is distinguished by the fact that the courtyard is surrounded on the three sides by the dwelling structures, and it is without a portico supported by pillars or columns as in the case of the palaces and some urban structures. It offers complete privacy for the inhabitants' courtyard activities as well as protection from the wind and the sun. In view of the socio-economic conditions of the investment required in the planning and size, this house-type seems to have been used exclusively by wealthy families. These were relatively spacious houses (Table 1). However, houses of this type in urban settings, not only protected the occupants from the dirt and noise of the street, but also utilized the limited urban space to the maximum.

Analogous structure to house no. (4) (Fig. 12a) was found in Structure K at ar-Risha (Helms, 1990: 117, Fig. 57). It has a symmetrical layout and consists of a courtyard with the entrance on the north-south axis. One wing of rooms/ cells was set into the north flank, bounded by a stone wall visible in the north and west. A rectangular wing or block of rooms (5x12m) lay on the north flank. The rectangular courtyard (12x19m) has a simple doorway in the south.

Regarding structure no. (19), one of the best constructed and most formal of all al-Hallabat houses (fig.12 b) which lies in front of the main façade of the Qasr, it represents an architectural type which fits into the much-debated category of palace, caravanserai (Khan), and castle (Qasr), with a possible early date (7th or early 8th century). Its large size ($633m^2$), the thickness of the walls (0.90m) made of large field stones, and location within the overall layout of the site, makes the building a potentially significant one. It has only 8 rooms and an extensive central courtyard ($256m^2$), with one huge room/ hall on the southeast side without any indication of internal partitions. Some walls were covered by marble and two plastered floor surfaces similar to that in the Qasr were uncovered. Also, the discovery of some glazed tesserae indicates that there was a mosaic pavement. It was probably roof-tiled as indicated by several roof tiles found around and inside the structure. The oval-shape structure located beside it might have been used for storage or animal pen.

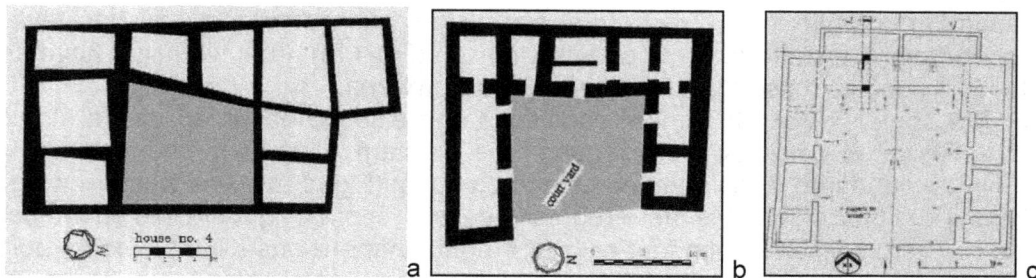

Figure 12. (a) plan of house no. (4) at al-Hallabat, (b) plan of house no. (19) at al-Hallabat, (c) plan of structure C at ar-Risha (Source: Helms, 1990: 85, Fig. 31).

An analogous example for structure no. (19) was also found in the Umayyad structure C (31.90mx31.30m) at ar-Risha (Fig. 12c). Helms believed that structure C is one of the best constructions and the most formal of all structures at the site, because of its square layout and relation with the mosque. The external stone base walls are made of also large field stones, well set and filled with smaller stones. The average width is between 0.80 and 1.20 meters. The building has only one entrance (3.38m wide) on the south side with series of rooms/ cells (4.2m on the east and west, and more irregularly from (4.2m to 3.8m) in the north. A good plaster floor was also found (Helms, 1990). It has also been suggested that while structure C at ar-Risha (Figs. 10c and 13c) may derive from an urban module, the houses at the same site stem partly from more rural origins, much as do the houses of today's recently settled bedouins and long-settled *fallahin* (Ahn, 2010).

The close layout and construction features of structure no. (19) and structure C is similar to the Khan at Qasr al-Hair al-Gharbi, which was originally made of mud brick on a stone base, dating back to about 727 AD. In the Khan at Qasr al-Hair al-Gharbi, however, there are six rooms of varying dimensions and a vestibule at the entrance side. Immediately to the right, while entering the courtyard, there is a block of masonry that served as the staircase to the roof (Creswell, 1989: 136). So probably structure no. (19) and structure C also had two floors, based on their relatively huge wall thicknesses. An analogous structure to house no. (19) can also be identified by some structures at Khirbet al-Askar, such as KS 3, KS 16, KS 17, KS 20, KS 23, and others (Kennedy, 2014: 104-05, Fig. 8 and Fig. 10). Other analogous parallel and related to ar-Risha structure C and structure no. (19) are found at Qasr al-Sawb (51x51m) at Dawqi/ ura (44x44m), at Jabal Seys/ Building F and G (34x35m), and perhaps also at structure KS10 at al-Risha.

All of these sites were possibly connected by ancient roads, with similar socio-economic conditions. All these examples also shared common features: a central square courtyard with only one entrance (doorway) on the central axis opening to the courtyard. However, the overall dimensions of the not cated Qasr al-Swab, which has an attached outer enclosure, are larger than at al-Hallabat and the other close parallels.

Finally, as mentioned before in all of these rural settlements, having similar socio-economic and political conditions, we can argue that the vibrancy of rural areas may have contributed to urban space becoming ruralized, with types of buildings already known in a rural context gradually introduced into the city (Polci, 2003: 101). At al-Hallabat rural area, some houses, such as house no. 1 and no. 3 can support this argument.

SUMMARY AND CONCLUDING REMARKS

The architecture of the so-called Umayyad desert palaces featured by a central square courtyard formed part of more extensive communities that engaged in a combination of agricultural activities and trade. These also demonstrate how the Umayyad patrons adapted and re-interpreted in a creative approach the military and domestic architectural traditions of late antiquity. Significantly, most of these buildings were abandoned soon after the fall of the Umayyad regime but they remain as evidence of the wealthy achievements of their dynasty.

However, the main architectural characteristic of early Umayyad domestic structures of the so-called rural 'Nomad Villages' and urban setting are also based on enclosing courtyard structures. Each of these Umayyad houses/ structures examined in this study has a courtyard, which functions as the nucleus and as an element of distribution, where most of the rooms have either direct or indirect access to the courtyard.

Al-Hallabat, the Umayyad so-called 'Nomad Village' houses represent and disseminate the nature and solidity of the high status of the socio-economic conditions of the early Umayyad Islamic occupation in Jordan. They might be considered as immediate predecessors of the wealthy rural houses of late antiquity, whatever form of their layout variety, scale, accessibility, and function arrangements.

However, it was possible to reveal a primary differentiation between the dwellings at al-Hallabat, ranging from relatively simple one-room structures to complex multi-family dwellings (Table 1, Fig. 5). Each type consists of at least one courtyard surrounded by a series of rooms or units in different arrangements and typological features, thus reflecting the socio-economic status of the householder. Three typological patterns were identified: nucleus house, courtyard house, and complex house. These types were then used to examine the position of the socio-economic conditions of al-Hallabat settlement in a tentative effort to map other analogous architectural examples based mainly on layout design and scale. In the simple nucleus house, the courtyard is alongside the house. The courtyard house was built around a courtyard from three sides to give greater privacy. The complex house is an expansion of the simple nucleus type, to which additional residential units and courtyards are attached according to the extended family needs.

For the identification of these three type patterns, a general classification rather than a detailed regional approach was adopted. Although it was occasionally dealing with the distinctive characteristics of these buildings of Bilad al-Sham, there are not yet enough systematically excavated houses in other parts of the region to make a workable regional typological pattern of the domestic Umayyad architecture. However, these three types/ patterns have provided helpful data for creating a general classification of the various socio-economic Umayyad house-types.

An important questioning of the veracity of the term "Nomad Village", even though it was used in this research, reveals that one should re-examine this term for al-Hallabat settlement. While, as shown, they confirm the continuation of the house types' used in the early Byzantine period in Bilad al-Sham, the study shows that meanwhile there are many similarities between al-Hallabat houses' layout and many other Umayyad houses in Jordan, still their main typology occurred in earlier examples found in the region.

What is of interest here is how the socio-economic conditions reflected at al-Hallabat typological patterns can be compared to other patterns found in more luxurious rural settlements, such as Um-el Jimal, and how these patterns evolved to the more distinctive forms that clearly define the attribution of these architectural arrangements to the Umayyad culture. The obvious mixture of urban and rural and rich and poor house architecture in al-Hallabat settlements might suggest that there were association and integration between urban and rural styles, rather than a departure to the countryside or to the city, where a more reciprocate exchange was in progress. Thus obscuring the distinction between the so-called 'Nomad Village' and town.

According to the comparative analytical study between Umayyad urban and rural Jordan, including other early Byzantine and Umayyad sites in the region, and al-Hallabat houses, the following conclusions can be drawn:

- The socio-economic conditions based on the general typological patterns at al-Hallabat layout seem to be very close to that at ar-Risha and Khirbet al-Askar, where their peak was also during the Umayyad period. Therefore, al-Hallabat and the building clusters at ar-Risha and Khirbet al-Askar need further research in relation to their location, space arrangement, typological pattern layout, landscape, and planning. Comparative studies between these sites can enrich our knowledge about the cultural significance for Umayyad Jordan and assist in establishing more accurate socio-economic typological patterns.

- There is an absence of the so-called 'bayt' unit in the residential structures of the houses presented in this study, with the exception of the houses at Amman citadel. None of the simple and direct relationship between tent (Bayt al-Sha'ar) and 'bayt' is evident at the early Islamic so-called 'Nomad Villages', such as ar-Risha and al-Hallabat until now. As previously suggested, it seems that the 'bayt' as a module is reserved primarily for palatial structures, as its main evidence is mostly found in the so-called desert palaces in Bilad al-Sham.

- None of the main architectural elements and related features of these houses, as also at al-Hallabat, can be considered to be totally new nor inventions of the period. The courtyard house is one of the oldest known architectural forms, particularly prevalent in the Mediterranean area and surrounding regions. The main layout of most of al-Hallabat's Umayyad modest houses is an inheritance from the early Byzantine and earlier periods. Many houses' layouts can be found during the early Byzantine period and continued to be used with slight changes at al-Hallabat during the Umayyad period. For example, the complex no. (1) at al-Hallabat has the analogous layout with the house at Umm el-Jimal (houses 7 and 8).

- At al-Hallabat agricultural settlement, the architecture of the complex and multi-unit living arrangement reflects a socio-economic aspect conducive to interdependence, with families functioning cooperatively in terms of land ownership, a division of labour and may be sharing working animals. The house layout arrangement is more advanced to what was exemplified at the early Islamic period farm in Naḥal Mitnan, as evident from the courtyard type house at al-Hallabat. This farmhouse type has only three dwelling units, each consisting of one to three rooms and a small courtyard.

Finally, it should be emphasized that the domestic architecture of only these examined sites cannot provide a sufficient basis for reaching comprehensive conclusions about the domestic Umayyad architecture of Jordan. Further detailed studies are required to reconstruct an exhaustive catalogue of housing in both urban and rural sites. Further archaeological research would reveal more convincing facts about how people lived and how they interacted socially and economically. The outcomes can subsequently be linked to the typological and architectural research. The outcome of such examination would also assist to develop a better understanding regarding the planning and the architectural design concepts of the so-called 'Nomad Villages' of the early Islamic period. Given the lack of Umayyad domestic architectural analytical studies, there is a vast need for further research within this area.

REFERENCES

Ahn, S. (2010). *The domestic architecture of Jordan-Palestine in the Early Islamic period: An archaeological approach* (Unpublished master thsis). The American University in Cairo, Cairo, Egypt.

Alhasanat, M.B., Wan, S.K.M., Aminuddin, W.H., and Addison, E. (2012). Spatial analysis of a historical phenomenon: Using GIS to demonstrate the strategic placement of Umayyad desert palaces. *GeoJournal, 77* (3), 343-359.

Almagro, A. (1987). Origins and repercussions of the architecture of the Umayyad Palace in Amman. In A. Hadidi (Ed.), *Studies in the History and Archaeology of Jordan: Third Edition* (181-192). Amman, Jordan: Department of Antiquities.

Almagro, A. (1992). Building patterns in Umayyad architecture in Jordan. In A. Hadidi (Ed.), *Studies in the History and Archaeology of Jordan: Fourth Edition* (351-356). Amman, Jordan: Department of Antiquities.

Almagro, A., and Arce, I. (2001). The Umayyad town planning of the Citadel of Amman. In A. Hadidi (Ed.), *Studies in the History and Archaeology of Jordan: Seventh Edition* (659-665). Amman, Jordan: Department of Antiquities.

Arce, I. (2007). Qasr Hallabat: Continuity and change from the Roman-Byzantine to the Umayyad period. In A. Hadidi (Ed.), *Studies in the History and Archaeology of Jordan: Ninth Edition* (325–344). Amman, Jordan: Department of Antiquities.

Bennett, C.M., and Northedge, A. (1977-1978). Excavations at the citadel, Amman, (1976): Second preliminary report. *Annual of the Department of Antiquities of Jordan, 22*, 172-179.

Bisheh, G. (1980). Excavation at Qasr Al-Hallabat. *Annual of the Department of Antiquities of Jordan, 24*, 69-79.

Bisheh, G. (1982). The second season of excavations at Hallabat, 1980. *Annual of the Department of Antiquities of Jordan, 26*, 133-143.

Bisheh, G. (1985). Qasr al-Hallabat: An Umayyad desert retreat or farm-land. In A. Hadidi (Ed.), *Studies on the History and Archaeology of Jordan: Second Edition* (263-265). Amman, Jordan: Department of Antiquities.

Bisheh, G. (1989). Hallabat (Qasr/ Qasr El). In D. Homes-Fredericq and J. Hennessy (Eds.), *Archaeology of Jordan. AKKADICA,7: Field Reports Surveys and Sites A-K*. Leuven, Belgium: Peeters.

Creswell, K. (1989). *A short account of early Muslim architecture*. Cairo, Egypt: The American University in Cairo Press.

De Vries, B. (1998). Umm el-Jimal: A frontier town and its landscape in Northern Jordan: Fieldwork 1972-1981. *Journal of Roman Archaeology, 1* (27), 91-127.

De Vries, B. (1981). The Umm el-Jmal project: 1972-1977. *Bulletin of the American Schools of Oriental Research, 244*, 53-72.

Gawlikowski, M. (1986). A residential area by the South Decumanus. In Fawzi Zayadine (Ed.), *Jerash Archaeological Project 1981-1983: Volume 1* (107-136). Amman, Jordan: Department of Antiquities.

Grabar, O. (1987). The formation of Islamic art. New Haven, C.T.: Yale University Press.

Grabar, O., Holod, R., Knustad, J., and Graber Trousdale, W. (1978). *City in the desert: Qasr al-Hayr East*. Cambridge, M.A.: Harvard University Press.

Genequand, D. (2006). Umayyad castles: The shift from late antique military architecture to early Islamic palatial building.. *History of Warfare Series, 35 (2)*, 3-25.

Ghrayib, R. (2003). The 2002 season at Qasr al Hallabat: A preliminary report. *Annual of the Department of Antiquities of Jordan, 47,* 65–74.

Ghrayib, R., and Ronza, M.E. (2007). Archaeological evidence of water distribution and quarrying activity at Qasr al-Hallabat. In A. Hadidi (Ed.), *Studies on the History and Archaeology of Jordan: Ninth Edition* (423–429). Amman, Jordan: Department of Antiquities.

Grube, E. (1978). What is Islamic architecture. In G. Michell (Ed.), *Architecture of Islamic World.* London, U.K.: Themes and Hudson.

Haddad, N. (2009). The architecture of Umayyad palaces in the Great Syria (Badiya): Analytical study in the particularity. *The Emirates Journal for Engineering Research, 14* (1),1 -8.

Hakim, B.S. (1986). Arabic-Islamic cities: Building and planning principles. London, U.K.: Kegan Paul International Ltd.

Harding, G.L. (1951). Excavations at the citadel, Amman *Annual of the Department of Antiquities of Jordan, 1,* 7-16.

Harding, G.L. (1984). The antiquities of Jordan. Jordan, Amman: Jordan Distribution Agency.

Helms, S. (1990). *Early Islamic architecture of the desert: A beduin station in Eastern Jordan.* Edinburgh, U.K.: Edinburgh University Press.

Haiman, M. (1995a). An early Islamic period farm at Nahal Mitnan in the Negev Highlands. *Atiqot, 26,* 1-13.

Haiman, M. (1995b). Agriculture and nomad-state relations in the Negev desert in the Byzantine and early Islamic periods. *Bulletin of the American Schools of Oriental Research, 297,* 29-53.

Hirschfeld, Y. (1995). *The Palestinian dwelling in the Roman-Byzantine period, SBF, 34.* Jerusalem: Franciscan Printing Press and Israel Exploration Society.

Jalboosh, F.Y. (2009). The architecture of the Umayyad Ssettlement around Qasr Al-Hallabat (Unpublished master thesis). Jordan Hashemite University, Amman, Jordan.

Kaptan, K. (2013). Early Islamic architecture and structural configurations. *International Journal of Architecture and Urban Development, 3* (2), 5-12.

Kennedy, D.L. (2014). 'Nomad villages' in north-eastern Jordan: From Roman Arabia to Umayyad Urdunn. *Arabian Archaeology and Epigraphy, 25* (1), 96–109.

Kennedy, D.L. (2000). *The Roman army in Jordan.* London, U.K.: The Council for British Research in the Levant.

Kennedy, D.L., and Riley, D.N. (1990). *Rome's desert frontier from the air.* London, U.K.: Batsford.

Magness J. (2003). *The archaeology of the early Islamic settlement in Palestine.* Winona Lake, IN: Eisenbrauns.

Magness J. (2004). Khirbet Abu Suwanna and Ein 'Aneva: Two early Islamic settlements on Palestine desert periphery. *Changing social identity with the spread of Islam: Archaeological perspectives.* Chicago, IL: The Oriental Institute of the University of Chicago.

McNicoll, A.W., Smith, R.H., and Hennessy, J.B. (1982). Pella in Jordan 1. *An interim report on the joint University of the Sydney and The College of Wooster Excavations at Pella 1979-1981.* Canberra, Australia: Australian National Gallery.

Northedge, A. (1992). Studies on Roman and Islamic Amman Volume 1: History, site and architecture. *British Academy Monographs in Archaeology (Book 3).* London, U.K.: Council for British Research in the Levant.

Petherbridge, G.T. (1978). Vernacular architecture: The house and society. In G. Michell (Ed.), *Architecture of the Islamic World* (176-208). London, U.K.: Thames and Hudson.

Petruccioli, A. (2007). After amnesia: Learning from the Islamic Mediterranean urban fabric .Bari, Italy: ICAR.

Piccirillo, M. (1984). The Umayyad churches of Jordan. *Annual of the Department of Antiquities of Jordan, 23,* 333-341.

Polci, B. (2003). Some aspects of the transformation of the Roman *domus* between late antiquity and the early middle ages. *Late Antique Archaeology, 1* (1), 79-109.

Toulan, N. (1980). Climatic considerations in the design of urban housing in Egypt. In G. Golany (Ed.), *Housing in Arid Lands: Design and Planning* (75-84). London, U.K.: The Architectural Press.

Urice, S.K. (1987). *Qasr Kharana in the Transjordan.* Durham, N.C.: American Schools of Oriental Research.

Walmsley, A. (2004). Archaeology and Islamic studies: The development of a relationship. In K. von Folsach, H. Thrane, and I. Thuesen (Eds.), *From Handaxe to Khan: Essays Presented to Peder*

Mortensen on the Occasion of His 70th Birthday (317-329). Aarhus, Denmark: Aarhus University Press.

Walmsley, A. (2000). Production, exchange, and regional trade in the Islamic east Mediterranean: Old structures, new systems?. In I.L. Hansen and C. Wickham (Eds.), The Long Eighth Century (265-343). Leiden, Netherlands: Brill Publishers.

Walmsley, A. (2007). Early Islamic Syria: An archaeological assessment. London, U.K.: Duckworth.

Walmsley, A. (2008). Households at Pella, Jordan: The domestic destruction deposits of the mid-eighth century. In L. Lavan, E. Swift, and T. Putzeys (Eds.), Objects in Context, Objects in Use: Material Spatiality in Late Antiquity (239-272). Leiden, Netherlands: Brill Publishers.

Watson, P.M. (1992). The Byzantine period: Byzantine domestic occupation in areas III and IV. In A. McNicoll, J. Hanbury-Tenison, J.B. Hennesy, T.F. Potts, R. Smith, A. Walmsley, and P. Watson (Eds.), Pella in Jordan 2: The Second Interim Report of the Joint University of Sydney and College of Wooster Excavations at Pella 1982-1985 (163-181). Sydney, Australia: Meditarch

Warren. J. (1978). Syria, Jordan, Lebanon. In G. Michel (Ed.), Architecture of the Islamic World. New York, N.Y.: Morrow.

AUTHORS

Naif A. Haddad
Associate Professor, Dr
Department of Conservation Science, Queen Rania's Faculty of Tourism and Heritage, Hashemite University, P.O. Box 3034, Amman 11181, Jordan.
naifh@hu.edu.jo

Fatima Y. Jalboosh
Private Practice, Ma
fatemah_y84@yahoo.com

Leen A. Fakhoury
Industrial Professor, MSc
School of Architecture and Built Environment, German Jordanian University, Amman,
lfakhoury962@gmail.com

Romel Ghrayib
Department of Antiquities, Zarqa, Jordan.
besan_g@yahoo.com

10

EVALUATION OF A POPULAR SHOPPING MALL BUILT TO ACCOMMODATE PREVIOUS STREET VENDORS IN DOWNTOWN PORTO ALEGRE

Naiana M. John, Antônio T. da Luz Reis, Márcia A. de Lima, Maria Cristina Dias Lay
Federal University of Rio Grande do Sul, Porto Alegre, Brazil.

*Corresponding Author's email address: naianaj@gmail.com, tarcisio@orion.ufrgs.br, malima.mgo@gmail.com, cristina.lay@ufrgs.br.

Abstract

The objective of this paper is to evaluate a popular shopping mall, named the CPC, focusing on its location, use, aesthetics and security. The global and local integration values of streets surrounding the CPC and the former Square XV Camelódromo are compared in order to evidence their level of accessibility. Concerning use, the following specific objectives are investigated: retailers' and CPC users' preference for a popular shopping mall type and CPC users' knowledge and shopping routines in the former Square XV Camelódromo; retailers' and CPC users' levels of satisfaction with the CPC, evaluations of the CPC building and shops location regarding sales, assessments of the existence of a food court and a bus terminal in the CPC building. In relation to aesthetics, retailers' and CPC users' evaluations of the internal appearance of the CPC, and retailers', CPC users' and architects' assessments of the external appearance of the CPC are revealed. Moreover, retailers' and CPC users' evaluations of security in the bus terminal in the CPC building and in downtown Porto Alegre are identified. Data gathering methods included archival surveys, physical measurements, behavioral observations, and questionnaires. Data were analyzed through non-parametric statistical tests and space syntax methods. Results corroborate findings of other studies regarding the need for movement of people and visibility of shops at ground level, in order to achieve a satisfactory sales performance. Moreover, the CPC external appearance tends to be differently evaluated among retailers, CPC users and architects.

Keywords: *Shopping Mall; Location; Use; Aesthetics; Security.*

INTRODUCTION

Informal street trade in Brazilian cities has been a concern to local governments due to its competition with nearby formal business and its obstruction of movement of people in the public spaces (Rigatti, 2003). Thus, in order to remove street vendors from downtown public spaces the implementation of popular shopping malls has been among the goals of redevelopments of urban areas in several Brazilian cities such as Porto Alegre, Belo Horizonte, Fortaleza and Joao Pessoa (Prefeitura Municipal de Joao Pessoa, 2010; Neves, Jayme, Zambellini, 2006; Dantas, 2005). Similarly, local governments of cities such as New York and Bogota have relocated street vendors to liberate public spaces for pedestrian circulation (p. ex. Donovan, 2008, 2002; Ya-Ting Liu, 2007; Devlin, 2006).

In order to achieve a successful outcome, such street vendors relocations and the consequent change of their condition to retailers in the new developments must consider shopping requirements (Zambellini, 2006). Shops need to be located in areas with a significant flow of people and/or activities that provide social gathering, as demand for products that are not basic necessities, such as toys and electronic equipments, depends on the opportunities created by meetings (Vargas, 2001; Rigatti, 2003). The less needed the products are, the more strategies are needed to encourage occasional purchases made by impulse. In this sense, a former study already emphasized the importance of shop location in order to reach as many customers as possible and increase sales (Lay and Oliveira, 2007). Early research shows that shops tend to be located along the most spatially integrated (accessible) streets (Hillier et al., 1993) and in streets with a high degree of connectivity to nearby streets (Hillier, 1999). Additionally, sales also depend on adequate visibility of the shops and on buildings configurations that do not divide the flow of customers (Vargas, 2001).

Nonetheless, distinct buildings configurations and locations were used in Brazilian cities to accommodate former street vendors. The local government of Belo Horizonte used an old brewery to relocate street vendors in the new Oiapoque Shopping (Figures 1 and 2). Shops were distributed on two floors and in the building courtyards and a public transport terminal was built to attract more people to the Shopping (Zambellini, 2006). The original building configuration and direct relationship with the public open spaces have been preserved, allowing pedestrians to visualize the shops.

Figure 1. View of Oiapoque Shopping. (Source: Google Earth, 2014).

Figure 2. Access to Oiapoque Shopping. (Source: Google Earth, 2014).

In others Brazilian cities, such as Porto Alegre and Joao Pessoa, new buildings were specifically designed to accommodate previous street vendors. Varadouro Shopping Center (Figure 3), located near a bus station in Joao Pessoa, consists of a two storey building with shops distributed on each floor (Prefeitura Municipal de Joao Pessoa, 2010). The configuration of a popular shopping mall in Porto Alegre (CPC – 'Centro Popular de Compras'), in turn, shows two blocks, three storeys high, connected by covered walkways crossing over Julio de Castilhos Avenue, with shops located on the second floor (Figures 4, 7,8 and 14).

Figure 3. Varadouro Shopping. (Source: Google Earth, 2011).

Figure 4. CPC. (Source: Celina de Pinho Barroso, 2009).

So far, however, no conclusive evidences from studies regarding the location, use, aesthetics, and security of buildings with different configurations used to accommodate previous street vendors, were found. This applies to the Popular Shopping Mall (CPC) in Porto Alegre, although press reports (Zero Hora, 2009a; Zero Hora, 2009b; Daroit, 2009) pointed out several problems, such as poorly built construction of buildings, tenants' arrears, small number of customers and occurrence of vandalism. Moreover, it is not known if the CPC distinct downtown location comparing to the former Square XV Camelódromo location (the early street vendors area; Figure 5) may have had any effect on movement of people and sales in the CPC.

Figure 5. Square XV Camelódromo – former main location of street vendors in downtown Porto Alegre, before the construction of the CPC. (Source: Celina de Pinho Barroso, 2009).

Therefore, the objective of this paper is to evaluate the popular shopping mall in downtown Porto Alegre, namely the CPC ('Centro Popular de Compras' - Popular Shopping Mall; Figure 4), focusing on its location, use, aesthetics and security. Regarding location, the global and local integration values of streets surrounding the CPC and the former Square XV Camelódromo (Figure 11) are compared in order to evidence their level of accessibility. Concerning use, the following specific objectives are investigated: retailers' and CPC users' preference for a popular shopping mall type and CPC users' knowledge and shopping routines in the former Square XV Camelódromo; retailers' and CPC users' levels of satisfaction with the CPC; retailers' and CPC users' evaluations of the CPC building and shops location regarding sales; retailers' and CPC users' assessments of the existence of a food court in the CPC; and retailers' and CPC users' evaluations of the incorporation of a bus terminal to the CPC building. In relation to aesthetics, retailers' and CPC users' assessments of the internal appearance of the CPC, and retailers', CPC users' and architects' appraisals of the external appearance of the CPC are revealed. Moreover, retailers' and CPC users' evaluations of security in the bus terminal, in the CPC building and in downtown Porto Alegre are identified.

METHODOLOGY
The object of this study, a popular shopping mall (CPC) located in a busy downtown area in Porto Alegre (Figure 6), characterized by services and commerce, was licensed by the Local Department of Industry and Commerce and inaugurated on February 9th, 2009. The CPC intended to accommodate 800 former street vendors previously working at the Square XV Camelódromo (Figure 6), also located in the city center of Porto Alegre. The building has an area of 20,000 m2 (twenty thousand square meters), constituted by two blocks (A side and B side) divided in three floors: in the ground floor (Figure 7) there is a bus terminal for more than 50 bus lines and by the time of this investigation there were 28 city bus lines on A side and 21

metropolitan lines on B side; in the second floor (Figure 8) there are shops (rented to retailers that were previous street vendors), a food court and other services; and the third floor is constituted by 216 parking spaces, a restaurant and the administration sector (Prefeitura Municipal de Porto Alegre, 2009).

Figure 6. Location of CPC and the former Square XV Camelódromo (main location of street vendors in downtown Porto Alegre, before construction of the CPC). (Source: Adapted from Google Earth by Celina de Pinho Barroso).

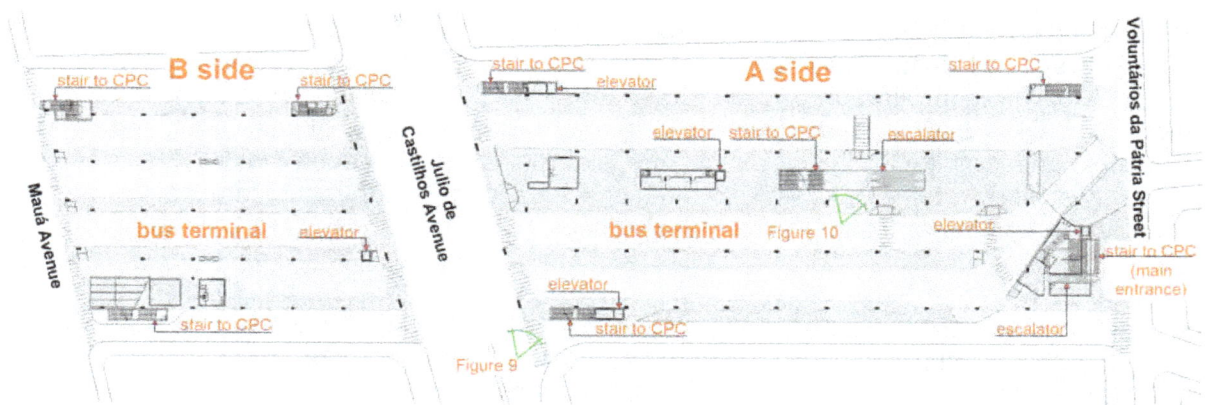

Figure 7. CPC ground floor plan. (Source: Adapted from Prefeitura Municipal de Porto Alegre, 2009).

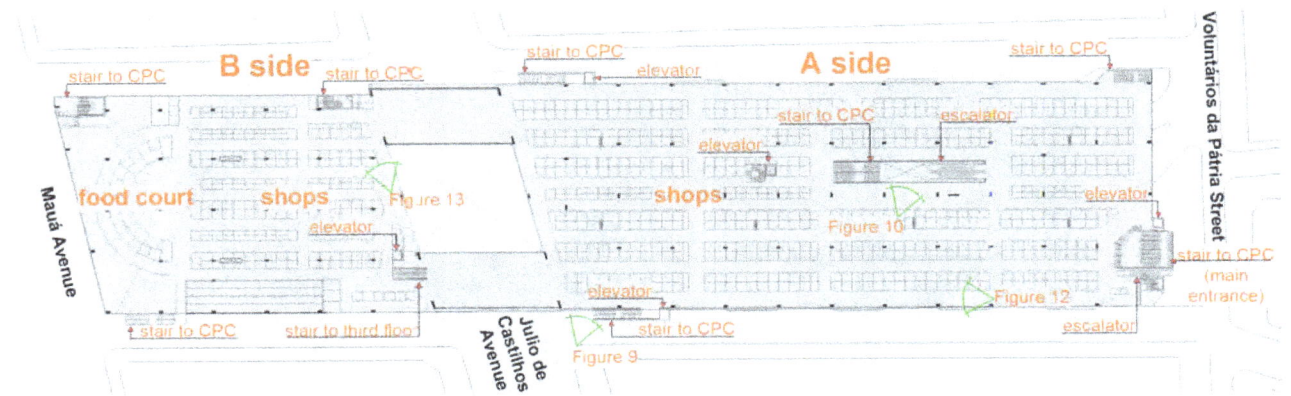

Figure 8. CPC second floor plan. (Source: Adapted from Prefeitura Municipal de Porto Alegre, 2009).

The access to the shops located on the second floor of the CPC occurs via stairs (Figures 7 and 9), escalators (Figures 7 and 10) and elevators; there are six staircases (five on A side and one on B side), two escalators on A side and five elevators (four on A side and one on B side; Figures 7 and 8). Universal accessibility for people with disabilities and reduced mobility is provided by the five elevators.

Figure 9. External access to the shops on A side. (Source: Author, 2009).

Figure 10. Access to the shops on A side by the bus terminal. (Source: Author, 2009).

Physical measurements, archival surveys, and behavioral observations were carried out, and questionnaires were administered to a sample of 128 respondents (Table 1), by the end of October and beginning of November 2009, nine months after the inauguration of CPC. Visits were made to the CPC during the morning of 22nd October in order to apply a pilot questionnaire. The final questionnaires were specific for each of the four groups (Table 1) and were applied during mornings and afternoons of the 29th and 30th October and morning of 5th November 2009 to CPC users (customers of shops and food court, and users of the bus terminal) and retailers. During this period questionnaires were also applied to architects at the Faculty of Architecture (UFRGS – Federal University of Rio Grande do Sul), at Porto Alegre City Council, and to city center users.

Table 1: Groups, subgroups and number of the respondents (Source: Author, 2012).

GROUPS	SUBGROUPS	NUMBER OF THE RESPONDENTS	TOTAL
Retailers (previous street vendors)	Retailers A side	17	31
	Retailers B side	14	
CPC users	Customers of shops	11	32
	Customers of food court	11	
	Users of the bus terminal	10	
Architects		33	33
City center users		32	32
TOTAL		128	128

Questionnaires including questions related to all the investigated variables were applied to retailers and CPC users; questionnaires with questions related to CPC external appearance were applied to architects and questionnaires with questions about shopping in the CPC and in the former Square XV Camelódromo were applied to city center users.

The external appearance of the CPC was evaluated considering the view from Julio de Castilhos Avenue - View 1 (Figure 14) and by comparison to a simulated view from the same observation point excluding the CPC building - View 2 (Figure 15), since these views reveal the major aesthetic impact produced by the CPC on the urban landscape. The quantitative data from the questionnaires were analyzed in SPSS (Statistical Package for the Social Sciences).

Nonparametric statistical tests such as Mann-Whitney U, Kruskal-Wallis and Spearman were performed with groups of respondents with, at least, 30 subjects (Table 1), which is considered an acceptable sample size to carry out nonparametric statistical tests [e.g. Reis, 1992, referring to Leedy's (1989) considerations].

Space syntax analysis was used in order to reveal location attributes through the global (considering all the other lines from each line in the urban system; in this case, the axial map o Porto Alegre) and local (considering only two steps, or two lines, from each line in the system) integration values of axial lines giving access to CPC and to the former Square XV Camelódromo. Higher integration values mean higher accessibility whereas lower integration values mean lower accessibility or higher segregation (Hillier and Hanson, 1984).

RESULTS AND DISCUSSION

Following the objectives, results are shown according to the main aspects of location, use, aesthetics and security.

Location - CPC and the former Square XV Camelódromo level of accessibility

In order to evidence their level of global accessibility, a comparison between the global integration values of streets giving direct access to CPC and to the former Square XV Camelódromo shows that these values are very similar, indicating a high level of accessibility in the urban system of Porto Alegre in both cases (Table 2; Figure 11). Moreover, although both are highly locally integrated, lines giving direct access to the CPC are slightly more locally integrated than those giving direct access to the former Square XV Camelódromo. This indicates that the first is a bit more accessible in downtown Porto Alegre than the second, suggesting that the location of CPC is very similar to the former Square XV Camelódromo in terms of being benefited by the movement of people (Table 2).

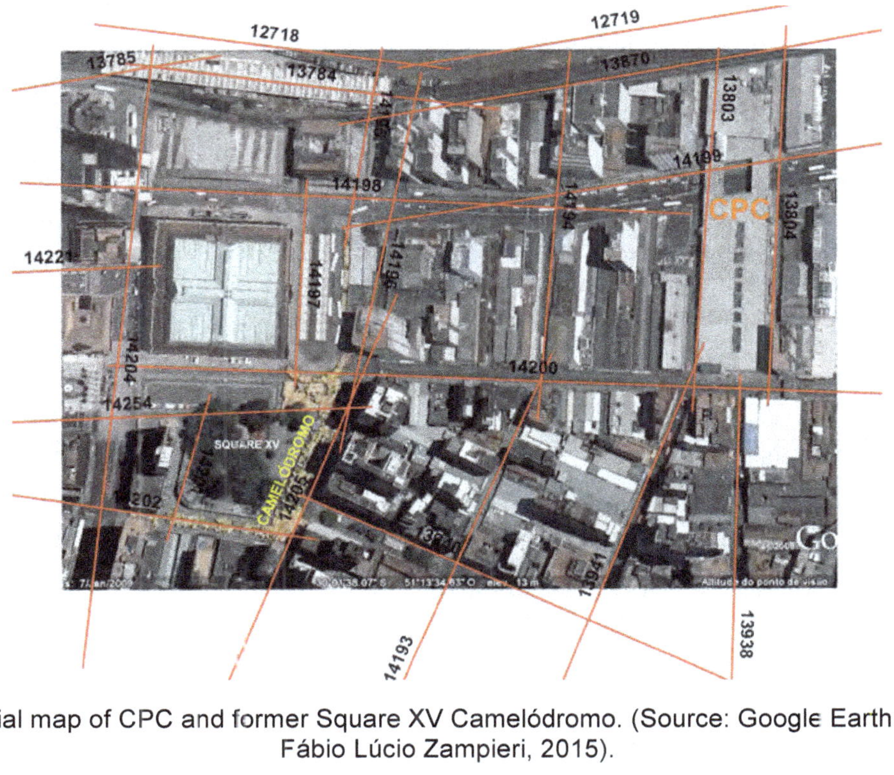

Figure 11. Axial map of CPC and former Square XV Camelódromo. (Source: Google Earth; axial map by Fábio Lúcio Zampieri, 2015).

Table 2: Global and local integration values of axial lines giving access to CPC and to former Square XV Camelódromo (Source: Author, 2015).

Line number	Street name	Global Integration	Local Integration R2
LINES GIVING DIRECT ACCESS TO CPC			
13803	Rui Barbosa Square	0,41752291	3,2369771
13804	Rui Barbosa Square (1)	0,41101205	2,8957148
13870	Mauá Avenue	0,41580322	4,5189767
13938	Senhor dos Passos Street	0,42479992	3,8564856
13941	Dr. Flores Street	0,4294745	4,120677
14199	Julio de Castilhos Avenue	0,4113082	4,605505
14200	Voluntários da Pátria Avenue	0,42542589	4,514534
Mean values		0,41933524	3,96412431
LINES GIVING DIRECT ACCESS TO FORMER SQUARE XV CAMELÓDROMO			
13940	Otavio Rocha Street	0,41462445	3,0307732
14196	Marechal Floriano Street	0,41774726	3,9872499
14197	Pereira Parobe Square	0,41047582	2,208734
14200	Voluntários da Pátria Avenue	0,42542589	4,514534
14201	Quinze de Novembro Square	0,40465194	2,3684211
14202	Jose Montaury Street	0,41892192	2,8350682
14205	Marechal Floriano Street	0,42969987	5,0405025
14254	Montevideo Square - Quinze de Novembro Square	0,4194065	5,3495517
Mean values		0,417619206	3,66685432

Use – Retailers' and CPC users' preference for a popular shopping mall type and CPC users' knowledge and shopping routines in the former Square XV Camelódromo

Considering the former Square XV Camelódromo, the CPC and other type of a popular shopping mall specified by the respondent, 92.96% of CPC users (27 out of 29) prefer the CPC (Table 3). The main reasons are the sheltered space of the building (40.7% - 11 out of 27), organization (33.3% - 9 out of 27) and security (18.5% - 5 out of 27). The clear preference of users for the CPC demonstrates the importance of having adequate space for shopping with protection from the weather, organized and safe. However, when questioned about types of popular shopping malls, many respondents were unable to think about any other type apart from the CPC itself.

Table 3: Preference for a popular shopping mall type including the CPC and the former Square XV Camelódromo (Source: Author, 2015).

	retailers A side	retailers B side	retailers (total)	customers of shops	customers of food court	bus terminal users	CPC users (total)
total	15 (100%)	14 (100%)	29 (100%)	10 (100%)	10 (100%)	9 (100%)	29 (100%)
Camelódromo	8 (53.3%)	4 (28.6%)	12 (40.9%)	0	0	1 (11.1%)	1 (3.7%)
CPC	6 (40.0%)	3 (21.4%)	9 (30.7%)	10(100%)	9 (90.0%)	8 (88.9%)	27(92.96)
Shop in square	1 (6.7%)	2 (14.3%)	3 (10.5%)	0	0	0	0
Square Ughini	0	1 (7.1%)	1 (3.55%)	0	0	0	0
CPC ground floor	0	2 (14.3%)	2 (7.15%)	0	0	0	0
On the street	0	2 (14.3%)	2 (7.15%)	0	0	0	0
None	0	0	0	0	1 (10%)	0	1 (3.33%)

Nonetheless, the expressive majority of CPC users (90.6% - 29 out of 32) knew the former Square XV Camelódromo and 65.6% (21 of 32) used to shop there (Tables 4 and 5). Therefore, most CPC users were able to compare the CPC and the Square XV Camelódromo.

Table 4:Knowledge about the Square XV Camelódromo (Source: Author, 2015).

	customers of shops	customers of food court	bus terminal users	CPC users (total)
total	11 (100%)	11 (100%)	10 (100%)	32 (100%)
yes	10 (90.9%)	10 (90.9%)	9 (90.0%)	29 (90.6%)
no	1 (9.1%)	1 (9.1%)	1 (10.0%)	3 (9 4%)

Table 5:Shopping in the former Square XV Camelódromo (Source: Author, 2015).

	customers of shops	customers of food court	bus terminal users	CPC users (total)
total	10 (100%)	10 (100%)	9 (100%)	29 (100%)
yes	8 (80.0%)	9 (90.0%)	4 (44.4%)	21 (71.4%)
no	2 (20.0%)	1 (10.0%)	5 (55.6%)	8 (23.5%)

\On the other hand, only 30.7% (9 out of 29) of the retailers (former street vendors at Square XV Camelódromo) prefer the CPC, while 40.9% (12 out of 29) prefer the former Square XV Camelódromo (Table 3) due to higher sales (75% - 9 out of 12), direct access from nearby streets (33.3% - 4 out of 12) and the presence of more people (33.3% - 4 out of 12). The main reasons for the preference for the CPC are security (66.7% - 6 out of 9), a better ambience (44.4% - 4 out of 9) and organization (33.3% - 3 out of 9). Therefore, the main reasons mentioned by retailers to justify preference for the former Square XV Camelódromo reveals the great importance for them of sales number, access and movement of people.

Use – Retailers' and CPC users' levels of satisfaction with the CPC
Regarding satisfaction with the CPC in general, a significant percentage of respondents (31.48% - 19 out of 63) considers it unsatisfactory or very unsatisfactory. Nonetheless, a statistically significant difference (Mann-Whitney U, sig.=.000) was found between retailers' and CPC users' degree of satisfaction with the CPC. Most retailers (56.30% - 17 out of 31) were dissatisfied or very dissatisfied, while the clear majority (90% - 29 out of 32) of CPC users (customers of shops, customers of food court and users of the bus terminal) evaluated the CPC as satisfactory or very satisfactory (Table 6). Hence, these results are consistent with retailers' and CPC users' preference for a popular shopping mall type.

Table 6: Degree of satisfaction with the CPC (SourceL Author, 2015).

GROUPS	v.sat.	sat.	n.n.	dis.	v.dis.	m.r.	Total
Retailers A side	0	6 (35,3%)	4 (23,5%)	3 (17,6%)	4 (23,5%)	-	17
Retailers B side	0	0	4 (28,6%)	6 (42,9%)	4 (28,6%)	-	14
Customers of shops	4 (36,4%)	7 (63,6%)	0	0	0	-	11
Customers of food court	2 (18,2%)	9 (81,8%)	0	0	0	-	11
Users of bus terminal	0	7 (70%)	1 (10%)	2 (20%)	0	-	10
Retailers - total	0	6 (17,65%)	8 (26,05%)	9 (30,25%)	8 (26,05%)	19,74	31
Users of CPC - total	6 (18,2%)	23 (71,8%)	1 (3,33%)	2 (6,66%)	0	43,88	32

Note. v.sat.= very satisfied; sat.= satisfied; n.n.= neither satisfied nor dissatisfied; dis.= dissatisfied; v.dis.= very dissatisfied; m.r.= mean rank values obtained through Mann-Whitney U non-parametric statistical test considering the group of retailers (31) and the group of CPC users (32).

Retailers' opinions reflect their knowledge about the CPC as a result of their daily experience and long stay in the shops during the day in different seasons of the year. However, retailers near the main entrance to the building on A side (Figure 8) are less dissatisfied with the CPC (41.1% dissatisfied or very dissatisfied - 7 out of 17) than the retailers on the B side (71.5% dissatisfied or very dissatisfied - 10 out of 14). The number and location of alternative entrances to the CPC (on the second floor of the building) generate differences in the number of customers in A and B sides, greater in A side as observed during visits to the CPC. So, this may justify the existing difference in the degree of satisfaction between retailers in A and B sides. While A side has eleven points of access (Figures 7 and 8), with escalator (Figure 8), staircases (Figures 7 and 8) and elevator, B side has only two, of which one access occurs via staircase when entering from Maua Avenue and the other access is via elevator adjacent to Julio de Castilhos Avenue (Figures 7 and 8).

Although CPC users are much more satisfied with CPC than the retailers (Table 6), according to the responses of 32 City Center users the number of these who usually buy in the CPC is 17 % lower than the number that used to buy in the former Square XV Camelódromo, revealing that the first has attracted fewer customers than the second used to attract. Nonetheless, the most mentioned reasons given by customers of shops in CPC are the low price of merchandise (52.3% - 11 out of 21) and the fact that they were passing by (33.3% - 7 out of 21), what supports the importance of location and people moving around a popular shopping to improve sales performance.

Use – Retailers' and CPC users' evaluations of the CPC building and shops location regarding sales

According to 81.3% (26 out of 32) of CPC users, the CPC building contributes to sales (Table 7) due to weather protection and comfort (76.9% - 20 out of 26), and safety (26, 9% - 7 out of 26), considered as improvements in comparison to the exposed open space of the former Square XV Camelódromo (Figure 5).

Table 7: Evaluation of the CPC building regarding sales (Source: Author, 2015).

contributes to sales	retailers A side	retailers B side	retailers (total)	customers of shops	customers of food court	users of bus terminal	users (total)
yes	9 (52.9%)	4 (28.6%)	13 (41.9%)	10(90.9%)	8 (72.7%)	8 (80%)	26(81.3%)
no	8 (47.1%)	10 (71.4%)	18 (58.1%)	1 (9.1%)	3 (27.3%)	2 (20%)	6 (18.8%)
total	17(100%)	14 (100%)	31 (100%)	11 (100%)	11 (100%)	10 (100%)	32 (100%)

On one hand, 58.1% (18 out of 31) of the retailers think that the CPC building does not contribute to sales, due to: the difficulty of access and low visibility of the shops located at the second floor (27.7% - 5 out of 18); the fact that shops would be better located at the street level (27.7% - 5 out of 18); and to the thermal discomfort (hot and humid, no cross ventilation) inside the building (16.6% - 3 out of 18). On the other hand, health problem caused by exposure to weather and thermal variations was the major complaint streets vendors' had regarding the Square XV Camelódromo (Machado, 2003). Nonetheless, despite the protection from the weather in the CPC building, according to many retailers, this feature does not replace the benefits of the being connected to the street regarding amount of people and sales. In total, 80.6% (25 out of 31) of retailers were dissatisfied or very dissatisfied with the sales in the CPC (Table 8). When asked to compare sales in CPC to sales in the former Square XV Camelódromo (Figure 5), 82.5% (18 out of 23) of all retailers and 100% of those located on B side said sales have decreased a lot in the CPC.

As expected, the correlation between the degree of retailers' satisfaction with the CPC and the degree of retailers' satisfaction with sales (Spearman, c.=.635, sig.=.000), confirms that satisfaction with sales affects retailers' satisfaction with the CPC. The main reasons associated with dissatisfaction with sales by retailers are: low movement of people (32% - 8 out of 25); high costs of rent and tax (28% - 7 out of 25); and poor access to shops on the second floor of the CPC (20% - 5 out of 25). This result is in accordance with the evaluations of Oiapoque Popular Shopping Mall in Belo Horizonte which shows that sales tend to be drastically reduced when shops are not on the ground floor (Zambellini, 2003).

Table 8: Retailers' satisfaction with sales in CPC and comparison to sales in the former Camelódromo (Source: Author, 2015).

DEGREE OF RETAILERS' SATISFACTION WITH SALES IN CPC						
	v.sat.	sat.	n.n.	dis.	v.dis.	Total
Retailers A side	0	3 (17,6%)	2 (11,8%)	5 (29,4%)	7 (41,2%)	17
Retailers B side	0	0	1 (7,1%)	4 (28,6%)	9 (64,3%)	14
Retailers - total	0	3 (8,8%)	3 (9,4%)	9 (29,0%)	16 (52,7%)	31(100%)
SALES IN CPC COMPARED TO SALES IN THE CAMELÓDROMO OF SQUARE XV:						
	g. incr.	ncr.	n.n.	decr.	g. decr.	Total
Retailers A side	0	2 (14,3%)	2 (14,3%)	1 (7,1%)	9 (64,3%)	14
Retailers B side	0	0	0	0	9 (100%)	9
Retailers - total	0	2 (7,15%)	2 (7,15%)	1 (3,55%)	18(82,5%)	23(100%)

Note. v.sat.= very satisfied; sat.= satisfied; n.n.= neither satisfied nor dissatisfied; dis.= dissatisfied; v.dis.= very dissatisfied; g. incr. = greatly increased; incr..= increased; n.n. = neither increased nor decreased; decr. = decreased; g. decr. = greatly decreased.

Additionally, a statistically significant difference (Mann-Whitney U, sig.=.000) was found between retailers' and CPC users' perception about the convenience of shops location at the second floor (Table 9). Among retailers, 83.3% (25 out of 31) believe that this configuration is unfavorable or very unfavorable to sales. The responses are also in line with newspaper news (Rodrigues, 2009) which revealed that, since the early days of CPC functioning, retailers were not satisfied with its infrastructure and asked for improvements to facilitate access by customers in order to boost sales. According to some retailers, sales decreased more than 50% since they moved from the former location in the streets to the CPC (Rodrigues, 2009), reinforcing previous results. The problem was acknowledged by Local Authority and as an alternative to improve sales in June 2009, the City Council and the CPC Administration took the initiative of changing CPC (Popular Shopping Mall) name to 'Porto Shopping - Camelódromo' (Zero Hora, 2009c) as this name was supposed to convey a better or refined CPC image. However this expectation of attracting more customers to the site did not occur, as revealed by the results of the CPC evaluation.

The percentage of dissatisfied CPC users (25% - 8 out of 32) with this configuration is substantially lower when compared to retailers' dissatisfaction; however, it is still significant the fact that a quarter of the CPC users who answered the questionnaire were dissatisfied with the location of shops on the second floor (Table 9).

Table 9: Evaluation of the location of shops on the second floor regarding sales (Source: Author, 2015).

	v.conv.	conv.	n.n.	inconv.	v.inconv	m.r.	Total
Retailers A side	0	1 (6,3%)	2 (12,5%)	4 (25%)	9 (56,3%)	-	16
Retailers B side	0	0	2 (14,3%)	2 (14,3%)	10 (71,4%)	-	14
Customers of shops	0	2 (18,2%)	8 (72,7%)	1 (9,1%)	0	-	11
Customers of food court	0	6 (54,5%)	2 (18,2%)	3 (27,3%)	0	-	11
Users of bus terminal	0	5 (50%)	1 (10%)	3 (30%)	1 (10%)	-	10
Retailers (total)	0	1 (3,3%)	4 (13,3%)	6 (20%)	19 (63,3%)	19,37	30
Users of CPC (total)	0	13 (40,6%)	11 (34,4%)	7 (21,9%)	1 (3,1%)	42,88	32

Note. v.conv.= very convenient ; conv.= convenient; n.n.= neither convenient nor unconvenient; inconv.= inconvenient; v.inconv.= very inconvenient; m.r.= mean rank values obtained through Mann-Whitney U non-parametric statistical test considering the group of retailers (30) and the group of CPC users (32).

Use – Retailers' and CPC users' evaluations of the existence of a food court in the CPC

The existence of the food court in the CPC was considered as positive or very positive by 77.65% of respondents (49 out of 63). Among the CPC users, 87.5% (28 out of 32) positively evaluate the existence of places for meals and snacks and 82.1% of these (23 out of 28) mentioned the convenience of these areas as the main reason. However, 43.3% of CPC users (13 out of 30) do not use the food court and 20% (6 out of 30) used it at most once a week. In turn, 72,7% (8 out of 11) of customers of food court shop at the CPC.

Among the retailers, 67.7% (21 out of 31) consider as positive or very positive the existence of a food court and 58.1% (18 out of 31) assume that these spaces contribute to sales. Moreover, 35.4% of retailers (11 out of 31) understand that the existence of a restaurant and snack bars facilitates their meals and 25.8% (8 out of 31) think that the food court attracts the public, while 12.9 % (4 out of 31) believe these places do not attract the public and do not affect sales. No further explanation was found for retailers in B side being much less satisfied (50% satisfied) with the nearby food court (Figure 8) than the retailers in A side (82.4% satisfied; Table 10). Nonetheless, the existence of correlation between levels of retailers' satisfaction with the existence of the food court and with the CPC (Spearman, c.=.393, sig.=.029), shows that retailers' satisfaction with the CPC is influenced by the existence of places for meals and snacks in the building.

Table 10:Evaluation of the existence of the food court in the CPC (Source: Author, 2015).

	v. positive	positive	n.n.	neg.	v. neg.	m.r.	Total
Retailers A side	7 (41.2%)	7 (41.2%)	3 (17.6%)	0	0	-	17
Retailers B side	2 (14.3%)	5 (35.7%)	6 (42.9%)	0	1 (7.1%)	-	14
Customers of shops	2 (18.2%)	8 (72.7%)	1 (9.1%)	0	0	-	11
Customers of food court	4 (36.4%)	7 (63.6%)	0	0	0	-	11
Users of bus terminal	1 (10%)	6 (60%)	3 (30%)	0	0	-	10
Retailers - total	9 (29%)	12 (38.7%)	9 (29%)	0	1 (3.2%)	30.47	31
Users of CPC - total	7 (21.9%)	21 (65.6%)	4 (12.5%)	0	0	33.48	32

Note. v. positive= very positive; n.n. = neither positive nor negative; neg. = negative; v.neg.= very negative; m.r.= mean rank values obtained through Mann-Whitney U non-parametric statistical test considering the group of retailers (31) and the group of CPC users (32).

Use – Retailers' and CPC users' evaluation of the incorporation of the bus terminal to the CPC building

The strategy of locating the bus terminal at the ground floor of the CPC building might be in accordance with the idea of using the movement of public transport users to promote sales in the CPC. However, it is evident that the terminal operates independently, with almost no visual integration (Figure 10) and with limited physical connection on B side (only one elevator and one stairs) to the shops at the second floor (Figures 7 and 8). Hence, the architectural features appear to explain why only 30% (3 out of 10) of users of the bus terminal purchase goods at the CPC.

In turn, a significant percentage of respondents (28.7% - 18 out of 63) are dissatisfied or very dissatisfied with the incorporation of the bus terminal to the CPC building (Table 6). However, there is a statistically significant difference (Mann-Whitney U, sig.=.002) between retailers' and users' levels of satisfaction with such integration. Among the retailers, 38.7% (12 out of 31) are dissatisfied with the integrated CPC building, and 56.6% (17 out of 31) consider that this integration does not contribute to sales. Moreover, 50% (6 out of 12) of dissatisfied retailers consider that it does not attract public and 41.6% (5 out of 12) believe that it would be better if the shops were located at street level. On the other hand, the main reason mentioned by satisfied retailers (35.5% - 11 out of 31) with the integrated CPC building was convenience of having the shops and the bus terminal in the same building (54.5% - 6 out of 11).

Among CPC users, 75% (24 out of 32) are satisfied or very satisfied with the incorporation of the bus terminal to the CPC building (Table 11), the main reason being the convenience of having the shops and the bus terminal in the same building (50% - 12 out of 24). Despite this positive evaluation, 53.1% of CPC users (17 out of 32) do not use the terminal and 9.4% (3 out of 32) use the bus terminal at most once a week.

Table 11:Evaluation of the incorporation of the bus terminal to the CPC building (Source: Author, 2015).

	v.sat.	sat.	n.n.	dis.	v.dis.	m.r.	Total
Retailers A side	1 (5.9%)	8 (47.1%)	5 (29.4%)	1 (5.9%)	2 (11.8%)	-	17
Retailers B side	0	2 (14.3%)	3 (21.4%)	6 (42.9%)	3 (21.4%)	-	14
Customers of shops	3 (27.3%)	4 (36.4%)	1 (9.1%)	3 (27.3%)	0	-	11
Customers of food court	1 (9.1%)	7 (63.6%)	0	2 (18.2%)	1 (9.1%)	-	11
Users of bus terminal	1 (10%)	8 (80%)	1 (10%)	0	0	-	10
Retailers - total	1 (3.2%)	10 (32.3%)	8 (25.8%)	7 (22.6%)	5 (16.1%)	25.27	31
Users of CPC - total	5 (15.6%)	19 (59.4%)	2 (6.3%)	5 (15.6%)	1 (3.1%)	38.52	32

Note. v.sat.= very satisfied; sat.= satisfied; n.n.= neither satisfied nor dissatisfied; dis.= dissatisfied; v.dis.= very dissatisfied; m.r.= mean rank values obtained through Mann-Whitney U non-parametric statistical test considering the group of retailers (31) and the group of CPC users (32).

The impact of the incorporation of the bus terminal to the CPC building on levels of satisfaction with the CPC building is supported by the existence of a correlation between such levels of satisfaction, either in the group of users (Spearman, c.=.378, sig.=.033) or in the group of retailers (Spearman, c.=.378, sig.=.036).

Aesthetics – Retailers' and CPC users' evaluations of the internal appearance of the CPC

The internal appearance of the CPC is negatively evaluated by 40.3% (25 out of 62) of the total respondents and no statistically significant difference between the evaluations of retailers and users was found. It was negatively evaluated by 51.6% (16 out of 31) of the retailers and by 29% (9 out of 31) of the total CPC users (Table 12).

Table 12: Evaluation of the internal appearance of the CPC (Source: Author, 2015).

	v. beautiful	beautiful	n.n.	ugly	v. ugly	m.r.	Total
Retailers A side	2 (11.8%)	2 (11.8%)	7 (41.2%)	4 (23.5%)	2 (11.8%)	-	17
Retailers B side	0	2 (14.3%)	2 (14.3%)	6 (42.9%)	4 (28.6%)	-	14
Customers of shops	2 (18.2%)	3 (27.3%)	4 (36.4%)	2 (18.2%)	0	-	11
Customers of food court	1 (9.1%)	2 (18.2%)	4 (36.4%)	2 (18.2%)	2 (18.2%)	-	11
Users of bus terminal	0	3 (33.3%)	3 (33.3%)	2 (22.2%)	1 (11.1%)	-	9
Retailers - total	2 (6.5%)	4 (12.9%)	9 (29%)	10 (32.3%)	6 (19.4%)	27.44	31
Users of CPC - total	3 (9.7%)	8 (25.8%)	11 (35.5%)	6 (19.4%)	3 (9.7%)	35.56	31

Note. v.beuatiful=very beautiful; n.n.= neither beautiful nor ugly; v.ugly=very ugly; m.r.= mean rank values obtained through Mann-Whitney U non-parametric statistical test considering the group of retailers (31) and the group of CPC users (31).

The main reasons given by retailers for a negative evaluation of the internal appearance of CPC are: poor construction and the unfinished look of the building (31.5% - 5 out of 16); the lack of color (31.5% - 5 out of 16); the ugly architectural appearance (31.5% - 5 out of 16) and the inadequate appearance for a shopping center (18.7% - 3 out of 16). The poor construction is related to leaks and spills, problems that delayed the opening of the CPC and remained till this investigation was carried out (Zero Hora, 2009a). The absence of color (also pointed out by 4 of the 9 users that evaluated the internal appearance as negative) refers to the fact that the CPC internal walls and roof are in grey of exposed concrete (Figures 12 and 13). Moreover, the inadequate look for a shopping center seems to be related to lack of paint or coating on the walls and floor, to the nonexistence of ceiling and to the exposed electrical and plumbing.

A statistically significant difference (Mann-Whitney U, sig.=.000) was found between the evaluations made by retailers and CPC users, specifically regarding the quality of construction and materials used in the CPC. While these aspects were positively evaluated by only 16.1% (5 of 31) of the retailers and negatively evaluated by 61.3% (19 of 31) of them, 67.7% (21 of 31) of CPC users evaluated them as satisfactory or very satisfactory (Table 13). The main reasons given by retailers for the negative evaluations are: spills and leaks (42.1% - 8 out of 19); bad materials (31.5% - 6 out of 19); and lack of finishing (15.7% - 3 out of 19), aspects that also tend to affect the evaluation of internal appearance, as already mentioned.

Figure 12. Internal view of CPC. (Source: Author, 2012).

Figure 13. Internal view of CPC. (Source: Author, 2012).

The main reasons given by CPC users who positively evaluate these aspects are the fact that the CPC was well built (19.0% - 4 out of 21) and its simplicity (14.3% - 3 out of 21).

According to these users, the ordinary materials used are appropriate since the CPC was built with public funds. So far, the differences between retailers and CPC users may be explained by the far greater familiarity and knowledge of CPC space by the retailers, comparing to CPC users.

Table 13: Evaluation of the quality of construction and materials used in CPC (Source: Author, 2015).

	v.sat.	sat.	n.n.	uns.	v.uns.	m.r.	Total
Retailers A side	0	3 (17.6%)	5 (29.4%)	4 (23.5%)	5 (29.4%)	-	17
Retailers B side	0	2 (14.3%)	2 (14.3%)	5 (35.7%)	5 (35.7%)	-	14
Customers of shops	1 (9.1%)	7 (63.6%)	2 (18.2%)	1 (9.1%)	0	-	11
Customers of food court	0	7 (63.6%)	3 (27.3%)	1 (9.1%)	0	-	11
Users of bus terminal	0	6 (66.7%)	0	2 (22.2%)	1 (11.1%)	-	9
Retailers - total	0	5 (16.1%)	7 (22.6%)	9 (29.0%)	10 (32.3%)	21.95	31
Users of CPC - total	1 (3.2%)	20 (64.5%)	5 (16.1%)	4 (12.9%)	1 (3.2%)	41.05	31

Note. v.sat .= very satisfactory; sat .= satisfactory; n.n. = neither satisfactory nor unsatisfactory; uns.= unsatisfactory; v.uns.=very unsatisfactory; m.r.= mean rank values obtained through Mann-Whitney U non-parametric statistical test considering the group of retailers (31) and the group of CPC users (31).

The existence of correlation (Spearman, c.=.297, sig.=.019) between retailers' and CPC users' levels of satisfaction with the CPC internal appearance and with the CPC indicates that satisfaction with the CPC is influenced by its internal appearance.

Aesthetics – Retailers' and CPC users' evaluations of the external appearance of the CPC
Regarding the external appearance of the CPC, 20.8% (20 out of 96) of the respondents (retailers, CPC users and architects) negatively evaluated View 1 (Figure 14), while 50% (48 out of 96) evaluated it as attractive or very attractive (Table 14). However, there is a statistically significant difference (Mann-Whitney U, sig.=.000) in the evaluation of View 1 among retailers, CPC users and architects. View 1 is evaluated as ugly or very ugly by 33.3% (11 out of 33) of the architects, by 25.8% (8 out of 31) of the retailers, and by only 1 (out of 32) user (Table 14). Therefore, View 1 is negatively evaluated by expressive proportions of architects and retailers.

View 2 (Figure 15; Table 14) is positively evaluated by 38.5% (37 out of 96) of respondents and negatively evaluated by 22.91% (22 out of 96) of them, with no statistically significant difference being found between retailers', users' and architects' evaluations. View 2 was negatively evaluated by 25.8% (8 out of 31) of retailers, by 24.2% (8 out of 33) of architects, and by 18.75% (6 out of 32) of CPC users.

Figure 14. View from Av Julio de Castillos with the CPC - View 1. (Source: Vanessa Dorneles, 2009).

Figure 15. View from Av Julio de Castillos without the CPC - View 2. (Source: Vanessa Dorneles, 2009).

When comparing the two views (Table 14), a significant percentage of the total respondents (33% - 32 out of 96) considers View 1 (with the CPC) uglier or much uglier than View 2 (without the CPC). A statistically significant difference (KW, Chi²=10.94, sig.=.004) was found in the evaluation of View 1 compared to View 2 between retailers, CPC users and architects, which shows that architects are the most dissatisfied with View 1 (Table 14), with 54.6% (18 out of 33) considering it uglier or much uglier than the View 2, followed by CPC users (25% - 8 ou of 32) and by retailers (19.4% - 6 of 31). The main reason for 33.3% (6 out of 18) of architects and 50% of CPC users (4 out of 8) is the visual barrier created by the CPC, which negatively affected the landscape.

Table 14: Evaluation of the external appearance of the CPC (Source:Author, 2015).

	very beautiful	beautiful	n.b.n.u.	ugly	very ugly	m.r.	Total
EVALUATION OF VIEW 1: View from Av Julio de Castillos with the CPC							
Retailers	0	15 (47.2%)	8 (26.6%)	5 (15.9%)	3 (10.1%)	45.65	31
CPC users	1 (3.1%)	26 (81.3%)	4 (12.5%)	1 (3.1%)	0	65.38	32
Architects	2 (6.1%)	4 (12.1%)	16 (48.5%)	9 (27.3%)	2(6.0%)	34.82	33
total	3 (3.1%)	45 (46.8%)	28 (29.2%)	15 (15.4%)	5 (5.36%)		96 (100%)
EVALUATION OF VIEW 2: View from Av Julio de Castillos without the CPC							
Retailers	3 (9.7%)	7 (22.6%)	13 (41.9%)	6 (19.4%)	2 (6.5%)	45.66	31
CPC users	4 (12.5%)	14 (43.8%)	8 (25%)	6 (18.8%)	0	56.91	32
Architects	0	9 (27.3%)	16 (48.5%)	7 (21.2%)	1(3%)	43.02	33
total	7 (7.4%)	30 (31.2%)	37 (38.4%)	19 (19.8%)	3 (3.1%)		96 (100%)
EVALUATION OF VIEW 1 COMPARING WITH VIEW 2							
	m.m. beautiful	m. beaut.	n.n.	uglier	m. uglier	m.r.	Total
Retailers	4 (12.9%)	17 (54.8%)	4 (12.9%)	2 (6.5%)	4 (12.9%)	54.6	31
CPC users	5 (15.6%)	16 (50%)	3 (9.4%)	7 (21.9%)	1 (3.1%)	55.28	32
Architects	1 (3%)	9 (27.3%)	5 (15.2%)	15 (45.5%)	3 (9.1%)	36.2	33
total	10 (10.5%)	42 (44%)	12 (12.5%)	24 (24.6%)	8 (8.3%)		96 (100%)

Note. n.b.n.ugly = neither beautiful nor ugly; m.m.beautiful = much more beautiful; m.beaut.= more beautiful; n.n.= neither more beautiful nor uglier; m.uglier= much uglier; m.r.= mean rank values obtained through Kruskal-Wallis non-parametric statistical test among the groups of retailers (31), CPC users (32) and architects (33).

Security – Retailers' and CPC users' evaluations of security in the bus terminal of CPC, in the CPC building and in downtown Porto Alegre

The percentage of respondents (retailers and users of the CPC) that considers the bus terminal (42.37% - 25 out of 59) and the CPC (22.58% - 14 out of 62) as unsafe or very unsafe clearly decreases in relation to those (79% - 49 out of 62) that consider downtown Porto Alegre as unsafe or very unsafe. No statistically significant difference between such evaluations by retailers and CPC users was found. Downtown Porto Alegre is considered unsafe or very unsafe by 80.6% of retailers (25 out of 31) and by 77.4% of CPC users (24 out of 31) (Table 15). The CPC is considered unsafe or very unsafe by 25.8% of the retailers (8 out of 31) and by 19.4% of users (6 out to 31). In relation to the bus terminal, the results indicate that 44.4% of retailers (12 out of 27) and 40.7% of CPC users (13 out of 32) regard it as an unsafe or a very unsafe place. Therefore, considering that the former Square XV Camelódromo was in downtown Porto Alegre, it seems that the security of users and retailers has been improved in the new location in the CPC. However, perception of insecurity in the CPC building and, mainly, in the bus terminal is not negligible. Results suggest that the main reason for the difference in the perception of security in the CPC building and in the bus terminal is the perception of darkness in this terminal.

Table 15: Perceived safety of retailers and users of the CPC (Source: Author, 2015).

	very safe	safe	n.n.	unsafe	very unsafe	m.r.	Total
EVALUATION OF SECURITY IN DOWNTOWN PORTO ALEGRE:							
Retailers	1 (3.2%)	3 (9.7%)	2 (6.5%)	15 (48.4%)	10 (32.3%)	30.97	31
CPC users	0	3 (9.7%)	4 (12.9%)	15 (48.4%)	9 (29.0%)	32.03	31
total	1 (1.6%)	6 (9.7%)	6 (9.7%)	30 (48.4%)	19 (30.6%)		62 (100%)
EVALUATION OF SECURITY IN THE CPC BUILDING							
Retailers	3 (9.7%)	15 (48.4%)	5 (16.1%)	5 (16.1%)	3 (9.7%)	30.37	31
CPC users	0	22 (71.0%)	3 (9.7%)	6 (19.4%)	0	32.63	31
total	3 (4.8%)	37 (59.7%)	8 (12.9%)	11 (17.7%)	3 (4.8%)		62 (100%)
EVALUATION OF SECURITY IN THE BUS TERMINAL OF CPC:							
Retailers	1 (3.7%)	7 (25.9%)	7 (25.9%)	9 (33.3%)	3 (11.1%)	28.81	27
CPC users	0	12 (37.5%)	7 (21.9%)	11 (34.4%)	2 (6.3%)	31.00	32
total	1 (1.8%)	19 (31.7%)	14 (23.9%)	20 (33.8%)	5 (8.7%)		59 (100%)

Note. n.n. = neither safe nor unsafe; m.r.= mean rank values obtained through Mann-Whitney U non-parametric statistical test considering the group of retailers (31) and the group of CPC users (32).

CONCLUSIONS

The results obtained from the evaluation of the CPC configuration confirm the findings of other studies (e.g. Vargas, 2001; Rigatti, 2003; Zambellini, 2006) regarding the need for movement of people and visibility of shops to achieve a satisfactory sales performance. This study further corroborates results obtained by Lay and Oliveira (2007) showing that the way people buy goods and services and the way sellers try to reach their potential customers seems to depend on the spatial configuration of an urban grid. Similarly to the location of shopping streets along the highly locally integrated streets, Lay and Oliveira (2007) showed that even low-income residents (predominantly illiterates or with poor education) understand that accessibility and consequent visibility is a location attribute needed to successfully perform income-generating activities that require visibility and tend to instinctively locate such activities along these streets.

Based on the opinion of retailers, it was evidenced the need for location of shops at ground level, since their location on the second floor is disconnected from the movement of people on the streets. This is supported by Zambellini's (2006) findings that shops on the second floor of the Oiapoque Popular Shopping Mall in Belo Horizonte did not achieve satisfactory sales performance. These findings also are in tune with Gehl's (2011; p.99) arguments presented in his book "Life between buildings: using public space": "In principle it is a bad idea to attempt to assemble activities by placing them above one another on different levels". Moreover, the fact that the CPC is located in an area slightly more beneficial to the movement of people than the former Square XV Camelódromo was not enough to prevent a reduction in CPC sales, due to its architectural configuration.

According to the results obtained, it can be inferred that the existence of the bus terminal in the CPC does not contribute to sales because the shops and the bus terminal operate independently, in different floors, with a small number of accesses on B side and very restricted visual connections that do not act as attractors to the shops on the second floor. As already mentioned, the success of sales, mainly those that are not basic necessities, depends on shops being visible and located in places with movement of people (Vargas, 2001).

The enclosed space of the CPC building is satisfactory for users due to the weather protection, comfort and safety, further contributing for shopping. The existence of the food court on the same floor of shops is a positive design characteristic, both for retailers and for users, by giving them the possibility of having meals without leaving the building, although the contribution of such food court to boost the sales is less evident.

The evaluation of CPC's overall appearance shows that internal appearance negatively affects retailers' and users' satisfaction with the CPC, mainly, due to poor quality of construction and materials used, the type of floors, walls and ceilings finishings and the lack of color, revealing the importance of these internal design attributes to user's satisfaction with a popular shopping mall. Complementary, the evaluation of CPC external appearance reveals a difference between architects and non-architects' aesthetics assessment, supporting results of some studies (e.g. Groat, 1982 apud Garling and Evans, 1991; Santos et al., 2011) but contrary to others (e.g. Reis et al., 2011) regarding the effect of type of college education on aesthetic evaluations. Nonetheless, the CPC creates a visual barrier for those in one of the busiest downtown streets and affect the urban aesthetics, which tend to be a negative design aspect for architects.

Considering that architecture and urban design deals with adequate relationships between elements in a building, between buildings and between buildings and open spaces, one can understand why the relationship of the CPC to other buildings and to the open spaces clearly tends to be negative for architects. Nonetheless, this clear tendency is not replicated for retailers and, especially, for users. While retailers do not have a clear negative or positive evaluation, the group of users has a clear tendency to positively evaluate the external appearance of the CPC. This may be, at least, partially explained by symbolic aesthetics. The CPC building, comparing to the former Square XV Camelódromo and even to the external appearance of buildings that constitute the built environment where many retailers and users live, may be associated with greater economic and social status, apart from its novelty (Santos et al., 2011).

Although perception of security in the bus terminal is negatively affected due to the lack of lighting, the enclosed space of the CPC building positively affects retailers' and users' perception of security in the building, when compared to perception of security in urban open spaces in downtown Porto Alegre.

Concluding, the results obtained in this study emphasize the need of understanding how buildings' location and configuration respond to users' requirements concerning use, aesthetics and security. Specifically, it shows the importance of assessing the impacts generated by current urban interventions such as popular shopping malls. The relocation of street vendors from downtown public open spaces in Brazilian cities to specific buildings, and the consequent liberation of public open space, seems to be satisfactory for city users. However, the configuration of popular shopping malls must conform to shopping requirements, taking advantage of the movement of people and creating meeting opportunities (e.g., Gehl, 2011; Ujang, 2014; Maimani, Salama, Fadli, 2014). Moreover, the design of a popular shopping mall must meet the needs and expectations of retailers and users as well as others, regarding its use, aesthetics and security.

REFERENCES

Al-Maimani, A., Salama, A., M. and Fadli, F. (2014). Exploring socio-spatial aspects of traditional souqs: the case of Souq Mutrah, Oman. *ArchNet-IJAR: International Journal of Architectural Research*, 8(1), 50-65.

Dantas, E.W.C. (2005). Apropriação do Espaço Público pelo Comércio Ambulante: Fortaleza-Ceará-Brasil em evidência - 1975-1995 (Appropriation of Public Space by Street Vendors: Fortaleza, Ceara, Brazil in evidence), *Scripta Nova Revista Eléctronica de Geografía y Ciencias Sociales*. Barcelona, vol. IX, n. 202. Retrieved from http://www.ub.es/geocrit/sn/sn-202.htm, Access Date, 26/04/2010.

Daroit, F. (2009, February 16). Baixo movimento faz com que varejistas se recusem a pagar o aluguel do camelódromo (Low movement causes retailers to refuse to pay the rent of shops in Camelódromo). *Zero Hora*. Retrieved from http://zerohora.clicrbs.com.br/zerohora/, Access Date, 01/10/2009.

Devlin, R. (2006). Illegibility, Uncertainty and the Management of Street Vending in New York City, *Breslauer Symposium*, University of California International and Area Studies, UC Berkeley. Retrieved from http://escholarship.org/uc /item/2dq8p606, Access Date, 11/02/2011.

Donovan, M. G. (2008). Informal Cities and the Contestation of Public Space: The Case of Bogotá's Street Vendors, 1988-2003, *Urban Studies*, 45:29-51. Retrieved from http://usj.sagepub.com/ content/45/1/29, Access Date, 09/02/2011.

Donovan, M. G. (2002). *Space Wars in Bogotá: the Recovery of Public Space and its Impact in Street Vendors* (Master's thesis). Massachusetts Institute of Technology, Indiana, USA.

Gärling, T., Evans, G. (1991). *Environment Cognition and Action: an Integrated Approach*. New York: Oxford University Press.

Gehl, J. (2011). *Life between buildings: using public space*. Washington: Island Press.

Hillier, B. (1999). Centrality as a Process: Accounting for Attraction Inequalities in Deformed Grids. *Urban Design International*, vol. 4, pp.107-127.

Hillier, B., Penn, A., Hanson, J., Grajewski, T., Xu, J. (1993). Natural Movement: Configuration and Attraction in Urban Pedestrian Movement. *Environment and Planning B: Planning and Design*, vol.20, pp.29-66.

Hillier, B., Hanson, J. (1984). *The Social Logic of Space*. Cambridge: Cambridge University Press.

Lay, M.C., Oliveira, C.H. (2007). An Analysis of Configuration, Location and Availability of Income-generating Activities in Social Housing. *Proceedings of the 6th International Space Syntax Symposium, İstanbul*, pp. 085-01-085-14.

Leedy, P.D. (1989). *Practical Research - Planning and Design*. New York, MacMillan.

Lima e Silva, P. (2009, January 23). Camelódromo recebe 49 linhas de ônibus (Camelódromo receives 49 bus lines). *Zero Hora*, p.48.

Machado, R. P., ROCHA, A. L. C. (2003). A rua como um estilo de vida: práticas cotidianas na ocupação do centro de Porto Alegre por vendedores ambulantes (The street as a lifestyle: everyday practices in the occupation of downtown Porto Alegre by street vendors), *Iluminuras, v(7)*. Retrieved from www.seer.ufrgs.br/index.php/ iluminuras/article/view/9156/5255, Access Date, 16/04/2010.

Neves, M., Jayme, J.G., Zambelli, P. (2006). Trabalho e cidade: os camelôs e a construção dos shoppings populares em Belo Horizonte (Labor and the city: the street vendors and the construction of the popular shopping malls in Belo Horizonte). *Proceedings of the encontro anual da ANPOCS*, pp. 199-199.

Prefeitura Municipal de João Pessoa (2010). Capital ganha Centro de Comércio do Varadouro nesta segunda-feira (Capital gets Varadouro Shopping Center on Monday), João Pessoa. Retrieved from http://www.joaopessoa.pb.gov.br/ noticias/?n=13427, Access Date, 06/07/2010.

Prefeitura Municipal de Porto Alegre. (2009). Projeto Viva o Centro (Viva o Centro Project). Retrieved from http://www2.portoalegre.rs.gov.br /vivaocentro/, Access Date, 01/10/2009.

Reis, A. (1992). *Mass housing design, user participation and satisfaction*. Thesis (Doctor in Philosophy, Post-graduate Research School, Oxford Brookes University, Oxford, England, 1992).

Reis, A., Biavatti, C., Pereira, M. L. (2011). Estética Urbana: uma análise através das ideias de ordem, estímulo visual, valor histórico e familiaridade (Urban aesthetics: an analysis through the ideas of order, visual stimulus, historical value and familiarity). *Ambiente Construído*, v. 11, n. 4, p. 185-204.

Rigatti, D. (2003). Camelôs, Flanelinhas e os Outros: Privatização de Espaços Públicos (Street vendors, People Who Look After the Parking Spaces and Others: Privatization of Public Spaces). *Paisagem Urbana*, São Paulo, Ensaios, n. 17, p. 41-67.

Rodrigues, E. (2009, March 07). Os dois lados do Camelódromo: Camelôs se queixam de queda nas vendas, ao contrário dos lojistas do Centro (The two sides of Camelódromo: Street vendors complain about falling sales, unlike downtown shopkeepers). *Diário Gaúcho*.

Santos, C. P., Barroso, C. P., Reis, A. T. L., Lay, M.C (2011). O Camelódromo e o Novo CPC – Centro Popular de Compras - Impactos Sobre O Espaço Urbano (The Camelódromo and the New CPC – Popular Shopping Mall – Impacts over the Urban Space). *Arquisur Revista*, v.1, p.104 - 115.

Ujang, N. (2014). Place meaning and significance of the traditional shopping district in the city centre of Kuala Lumpur, Malaysia. *ArchNet-International Journal of Architectural Research*, 8(1), 66-77.

Vargas, H. C (2001). *Espaço terciário: o lugar, a arquitetura e a imagem do comércio* (Tertiary space: the place, the architecture and the image of shopping). São Paulo: Editora SENAC São Paulo.

Ya-Ting Liu (2007). *A Right to Vend: New Policy Framework for Fostering Street Based Entrepreneurs in New York City* (Master's thesis). Massachusetts Institute of Technology, Berkeley, California, USA.

Zambellini, P. H. L (2006). *O trabalho informal dos camelôs na região central de Belo Horizonte e a transferência para os Shoppings Populares* (The unofficial work of the street vendors in the central region of Belo Horizonte and the transfer to the Popular Shopping Malls). Dissertation (Master in Social Sciences), Pontifícia Universidade Católica de Minas Gerais, Belo Horizonte.

Zero Hora. (2009a, February 11). Pelas Ruas: recém-inaugurado, Camelódromo da Capital apresenta goteiras (On the Streets: newly inaugurated, Camelódromo in the Capital presents gutters. Retrieved from http://zerohora.clicrbs. com.br/zerohora/, Access Date, 01/10/2009.

Zero Hora. (2009b, February 12). Vandalismo atinge o Camelódromo (Vandalism reaches the Camelódromo). Retrieved from http://zerohora.clicrbs.com.br/zerohora/, Access Date, 01/10/2009.

Zero Hora. (2009c, June 23). Camelódromo de Porto Alegre terá mudanças visuais: inadimplência de locatários chega a 30% (Camelódromo of Porto Alegre will have visual changes: default tenants reaches 30%). Retrieved from http://zerohora.clicrbs.com.br/zerohora/, Access Date, 01/10/2009.

AUTHORS

Naiana Maura John
Master in Urban and Regional Planning
PROPUR (Post Graduate Program in Urban and Regional Planning)
Federal University of Rio Grande do Sul, Porto Alegre, Brazil
Email address: naianaj@gmail.com

Antônio Tarcísio da Luz Reis
Professor, Ph. D.
Faculty of Architecture - PROPUR (Post Graduate Program in Urban and Regional Planning) Federal University of Rio Grande do Sul, Porto Alegre, Brazil
Email address: tarcisio@orion.ufrgs.br

Márcia Azevedo de Lima
Doctoral Researcher
PROPUR (Post Graduate Program in Urban and Regional Planning)
Federal University of Rio Grande do Sul, Porto Alegre, Brazil
Email address: malima.mgo@gmail.com

Maria Cristina Dias Lay
Professor, Ph. D.
Faculty of Architecture - PROPUR (Post Graduate Program in Urban and Regional Planning) Federal University of Rio Grande do Sul, Porto Alegre, Brazil
Email address: cristina.lay@ufrgs.br

11

CONSTRUCTION TECHNIQUES ON THE BOSPORUS REGION

Fatih Yazicioglu*
Istanbul Technical University, Istanbul 34367, Turkey

*fyazicioglu@gmail.com, yaziciogluf@itu.edu.tr

....

Abstract

The Bosporus is a unique place with its traditional architecture. The construction of new buildings along the Bosporus coast line are forbidden and only reconstructions of demolished original historic buildings are possible. Since 1984 legal authorities act different legislations about these reconstructions. Until 2005 the legislation gives permission for the usage of any structural material unless the overall appearance is the same as the original building. But after 2005 the legislation changed and the usage of original structural material became mandatory. The underlying reason of this is to protect the originality and authenticity of the Bosporus buildings both with their appearances and construction techniques. But the practical output became promoting the constructers for making the reconstructions without following the approved reconstruction projects or changing the reconstructions afterwards. In this paper structural system, external wall, window, and roof alternatives for reconstructions are compared and contrasted by evaluating the performance requirements. Examples of inappropriate reconstructions are documented and a tentative proposal is made for the reconstructions of the Bosporus region.

Keywords: *Construction Techniques; Structure; Building Elements; Residential Housing; Bosporus*

INTRODUCTION

Istanbul is divided into two parts between Asia and Europe by the waterway Bosporus. Bosporus connects Black Sea with the Sea of Marmara and is about 30km long. Its coastal region consists of several villages where authentic residential Istanbul buildings exist. There are three main types of buildings in the villages of the Bosporus; monumental palaces, "yalı" buildings, and mansions. Monumental palaces were the residential buildings of the Ottoman dynastic family and, yalı and mansions were other residential buildings for nobles and common people. The main difference between a yalı and a mansion lies within their locations. Yalı buildings were constructed on the coast of Bosporus by the sea, whereas the mansions on inner parts (Sakoglu, 2012). Residential building typology of Bosporus has gained it a unique building texture.

In order to protect the unique building texture of Bosporus a special law, dated to 1984, has been acted through which new building constructions are forbidden (Grand National Assembly, 1984). It is only possible to make restorations for old buildings, and to reconstruct a building only if it can be proved that there used to be an original historical building in that place (Kanadoglu, 2009). Two bodies were authorized over Bosporus by the law; the first one is the "Cultural and Natural Heritage Preservation Board". Its functions are; to assess the buildings of Bosporus, to categorize them into groups, and to evaluate the demands for reconstructions in means of architectural restoration or reconstruction projects. The second body is the "Directorate of Bosporus Housing" and its function is to control the restorations, and the reconstructions (Istanbul Metropolitan Municipality, 2014).

On Bosporus region, apart from the monumental palaces, almost all of the residential buildings initially constructed of wood (Hisar, 2012). Since 1980s several of them were reconstructed with different construction techniques. These reconstructions are a debate for a long time and there are mainly two different aspects: In the first aspect, it is claimed that these houses are authentic as a whole with their construction techniques. Whereas, in the second aspect it is claimed that; the overall characters of the facades and of the districts are authentic and so the construction techniques of the houses may be changed. Until 2005 the second aspect had been practically used, after then, the first aspect was started to be used and; the usage of the original construction techniques, especially the original structural system, has been started to be compulsorily demanded by the related legal authorities (Kanadoglu, 2009). As a result, different construction techniques in terms of structures were used for the reconstructions on the Bosporus. The reconstructions realized according to the first aspect use wood as the structural component, and reinforced concrete is used with the second aspect. The significant thing about these reconstructions is that the difference of the structures cannot be distinguished from outside appearance. Nevertheless, these different types of structures have different positive and negative properties and discussing the subject only from conservation perspective is not sufficient. For example, wooden structures are restricted in the "İstanbul Building Construction Code", because of the absence of a specific code related to wooden structures (Istanbul Metropolitan Municipality, 2007). Most of the reconstructed buildings of the Bosporus do not meet these restrictions and this conflict should be discussed and corrected. On the other hand several original details of the reconstructed buildings with the first aspect, have been changed for improving the performance of them. Double glazed windows, and thermally insulated external walls are examples for these kinds of changes about the details of the buildings. Although the reconstructions on the Bosporus are specially treated by the law, because of their authentic worth, it is also expected from them to comply with several legislations, like the legislation of energy efficiency in buildings (Ministry, 2008). In order to comply with this legislation the original façade details of the buildings should be changed and these changes affect the authenticity of the buildings no matter the original structural material was used or not. Sustainability is another critical issue about these buildings. Although wooden structures seem to be more environment friendly, the usage of excessive insulation materials lower the structures sustainability (Berge, 2009). Furthermore, the operational energy costs of the system, which is the energy consumed during the in-use phase of a building's life, in terms of CO_2e is higher than the reinforced concrete structured buildings (Yazicioglu, 2012), (Cole and Kernan, 1996). Hence sustainability related issues should also be studied in detail while discussing the reconstructions of the Bosporus. These examples demonstrate that, the reconstructions on the Bosporus should not only be studied about their authentic properties, but also the detail design properties of them should be analysed, and the decisions about the buildings should be given with an integrated evaluation.

In the paper, examples of original and reconstructed residential buildings of Bosporus region will be analysed, giving an emphasis on the construction techniques. The legislations, related with the conservation of the buildings and performance requirements will be compared and contrasted. At the end a tentative proposal about the reconstructions on the Bosporus region will be given.

METHOD

The method adopted for this research consists of two main parts. The first part is comparing and contrasting the selected systems of the reconstructions. Four systems; the structural system, external wall, window, and roof were selected to be compared and contrasted. The reasons for selecting these systems are as follows: Having a structural system is the primary criteria for the existence of a building as it gives the building its shape and it resists to all loads coming to the building (Engel, 1981). Thus, the structural system is the first building element selected to be examined. Having an envelope is also a primary criteria for having a good performing building as it determines the building's structural stability, climate control, and degree of energy performance

(Lovell, 2010). Thus the 3 elements of the building envelope; the external wall, the window, and the roof are the other 3 systems selected to be examined.

The comparing and contrasting was realised by evaluating the performance of the alternatives. Considering the most important user requirements expected from the systems, a total of 7 performances were chosen to be compared and contrasted (Rich, 1999). 5 of these are evaluated with respect to the detailed case and literature reviews. These are structural strength and stability, acoustics, water & moisture, fire, and durability related performances. In order to compare and contrast these 5 performances of the systems key detail drawings have been generated considering the analysed cases and literature.

The other 2 performances, which are thermal and sustainability related, are mathematically calculated, compared and contrasted. For thermal performance; EN 832 standard has been followed for walls, and ISO 10077-2 standard has been followed for windows, and U-Values have been calculated (Standards British, 2000, Standards International, 2012). U-Value is the overall heat transfer co-efficient, in other words it shows the mathematical value of the heat loss in a building element such as a wall, floor or roof (Bougdah, 2009). Thus the calculation of it made it possible to understand thermal performance of the systems. The U-Value also shows the success of a detail about the operational heating energy losses of the systems. Operational energy usage shows the success of the building about sustainability concerns because it shows the CO_2 footprint of it. CO_2 footprint is the total amount of harmful emission that has been given to the atmosphere during the life of a building (Cook, 2011). Furthermore for evaluating the sustainability related performance of the systems embodied CO_2 have also been considered. The embodied CO_2 is the total amount of harmful emission in the production step of a material (Cleveland& Morris, 2009).

The second part is the evaluation of the cases about the selected systems. The cases are selected by making site visits and the most significant cases are used as examples in the paper. The common feature of the selected cases are to have modern construction techniques and materials together with the original and authentic details. These cases are critically analysed and the reasons why modern construction techniques and materials are used in the reconstructions are tried to be understood. Finally a tentative proposal is made for the reconstructions on the Bosporus.

COMPARATIVE ANALYSIS OF ORIGINAL DETAILS WITH RECONSTRUCTION ALTERNATIVES

In order to analyse the original buildings of Bosporus four different building sub-systems are going to be discussed. The first one is the structural system. The structural system is a critical element of the reconstructions on the Bosporus as the usage of original structural system material or a contemporary one is one of the major debates (Cultural and Natural Heritage Preservation Board, 2005). Also it gives the building its shape and protects this shape under the effect of loads which make the structural system one of the most important elements of all buildings (Turkcu, 2003). Thus the structural system is chosen to be the first sub-system that is going to be analysed. The external wall is the vertical opaque component of the external envelope. Most of the performance requirements expected from the buildings are satisfied by the external walls as they form the largest area of the external building envelope (Brock, 2005). Thus the external wall system is chosen to be the second sub-system that is going to be analysed. Together with the external walls the window systems are also the vertical components of the external envelope. In order to satisfy the transparency needs of the external envelope the windows are the most critical points. Meanwhile, windows are systems where several leakages occur, like thermal leakage (Carmody, et al. 2000). Thus the windows are the third sub-system that is chosen to be analysed. The roofs are the horizontal/semi horizontal parts of the external envelope. It is directly and critically under the effect of atmospheric conditions. It is also a challenging part of the building for the architects to transform it into a residential space (Harrison, 2000). Thus the roofs are the final sub-system chosen to be analysed. In this part of the paper these systems are going to be analysed, considering mainly the methods/approaches used in the restorations/reconstructions of Bosporus district.

Structural system

On the Bosporus district originally stone is used in the basements and foundations (substructure), and wood in the upper floors and the roofs (superstructure). The superstructures are typically consist of two or three storeys and an attic. This character of the original buildings limited the spans to be four meters at most. The plan schemas and the facade characters of the original buildings were shaped by this very basic structural limitation. The substructure is the critical end point of the structural system from where the loads are transferred to the ground. Originally because of the sloped topography most of the residential buildings on the Bosporus district used to have a partial basement floor. The typical structural system details of the original buildings can be seen in figure 1.

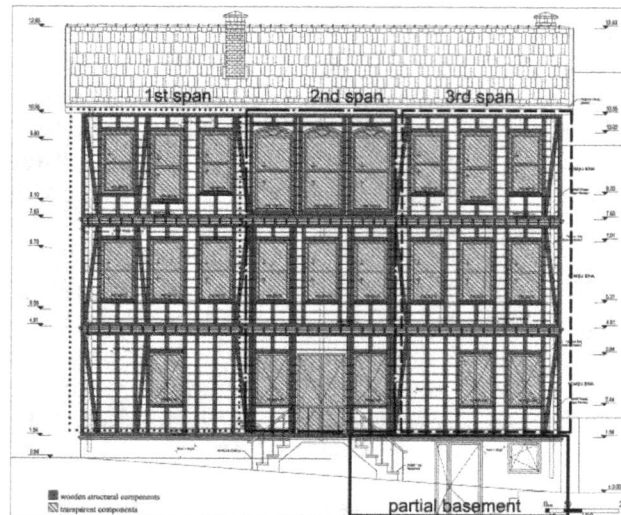

Figure 1. Typical facade drawing, showing three different spans and the place of the partial basement (Source: Author).

In the reconstructions, the substructure of the buildings are made of reinforced concrete. Both the legislation related with the substructure and the principles determined by the "Cultural and Natural Heritage Preservation Board" obligates this (Cultural and Natural Heritage Preservation Board, 2005). And the original partial basement is transformed to be a full scale basement because of the same reasons. But about the superstructure a fuzzy situation exists. Until 2005 any structural material was accepted to be used in the reconstructions. Hence mainly reinforced concrete (R.C.) and steel structures were used until then. The reasons of that is; firstly because of the needs of the contemporary living style, larger spans are expected by the users. And R.C. and steel give the possibility to pass larger spans. Secondly, R.C. and steel structures perform better about the sound, fire, and durability related performances.

In table 1 performances related with wooden, R.C., and steel structural materials are evaluated. If masonry and R.C. substructures are compared it is found out that R.C. substructures are slightly better than the masonry structures. The performance related with impacts are satisfied better with R.C. substructures because they are more homogenous and instead of adhering single large pieces, small pieces are adhered to each other. Thus it is harder to harm R.C. substructures by impacts. The same thing is also valid about the pressured underground water. As larger spaces exist between stones used in masonry structure, it acts worse about pressured underground water.

If wooden, R.C., and steel superstructures are compared it is found out that R.C. superstructures are slightly better than the other two. As R.C. is denser than the other two, performance related with airborne sound is the first performance which R.C. overcomes. R.C. also performs better about all 5 performances related with fire. Both wooden and steel superstructures will perform badly if they are not treated specially against fire, whereas R.C. is naturally performing well under fire. R.C. and steel performs similar about durability related performances and they are performing slightly better than wooden structures, especially about the resistance to biologicals.

Table 1. Evaluation of the structural system materials which may be used in the reconstructions.

		masonry substr.	R.C. substr.	wooden superstr.	R.C. superstr.	steel superstr.
Performance related with structural strength and stability	own weight	+	+	+	+	+
	earthquake	+	+	+	+	+
	impacts	-	+	-	+	-
	wind	+	+	+	+	+
	water pressure	-	+	/	/	/
	soil pressure	+	+	/	/	/
Performance related with acoustics	impact sound	-	-	-	-	-
	air born sound	+	+	-	+	-
Performance related with fire	inflammable	+	+	-	+	-
	continuity of str. safety	+	+	-	+	-
	harmful gases emission	+	+	-	+	-
	keep the fire in its place	+	+	-	+	-
	keep the smoke in its place	+	+	-	+	-
Performance related with durability	resistance to chemicals	+	+	+	+	+
	resistance to biologicals	+	+	-	+	+
	res.to mechanical move.	-	+	+	+	+
		12/16	15/16	5/14	13/14	6/14

External Wall

The external walls of the original buildings were wooden stud walls which were also acting as the vertical load bearing component. In table 2 performances related with original wooden, reconstruction alternative wooden, and blockwork structured walls are evaluated. The typical details of the original external walls can be seen in figure 2 left. Mainly these external walls were typically 16 cm thick, which consist of 2 cm of wooden siding outside 12 cm of air cavity in between the studs and 2 cm of wooden siding (+plaster) inside. The U-value of this detail has been calculated and is approximately 0.73 W/m2K.

There are basically 2 different types of details for reconstructions on the Bosporus district. The first detail is used for reconstructions with wooden stud walls. The typical external wall details of this type can be seen in figure 2 middle. Mainly these external walls are similar with the original one, the only difference is the thermal insulation material used in the air cavity. The resulting U-value of this type of external wall has been calculated, and is approximately 0.51 W/m2K. The second detail is for reconstructions with blockwork walls. The typical external wall details of this type can be seen in figure 2 right. These external walls are typically about 30 cm thick, which consist of 2 cm wooden siding, 4 cm of thermal insulation, 2 cm of external plaster, 20 cm of solid core, and 2 cm of internal plaster. The resulting U value of the external wall has been calculated, and is approximately 0.46 W/m2K. When thermal performance of wooden and blockwork walls are compared it is found out that blockwork walls are slightly better than the other two wooden structures.

If a building's life cycle of 30 years, which is heated by a natural gas burning boiler, is calculated for each external wall detail it is found out in a previous study that the CO2 footprint of the first detail is larger than the other two, and the last detail (R.C. structure) is the smallest of the alternatives (Yazicioglu, 2012). The blockwork wall alternative is also overcoming the other two about the performance related with fire and durability.

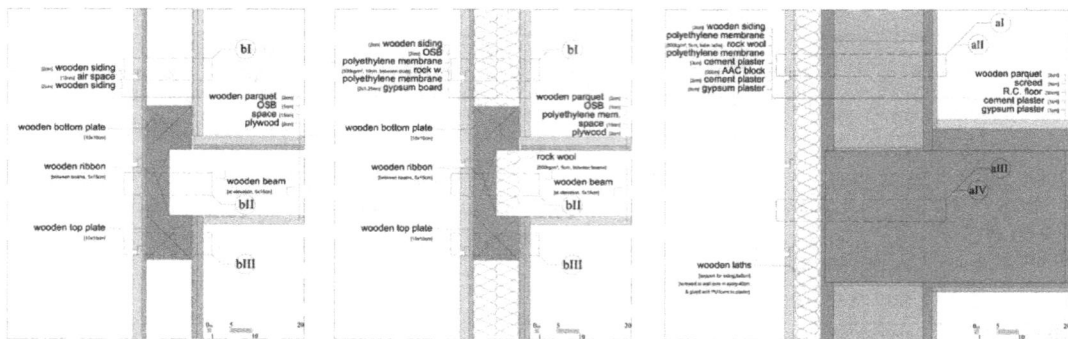

Figure 2. (Left) section of the original buildings external wall details; (middle) section of the external wall of the reconstructions with wooden stud walls; (right) section of the external wall of the reconstructions with blockwork walls (Source: Author).

Table 2. Evaluation of the original external wall with reconstruction alternatives.

		original detail	wooden wall	blockwork
Performance related with structural strength and stability	own weight	+	+	+
	earthquake	+	+	+
	impacts	+	+	+
	wind	-	+	+
performance related with thermal issues	low heat transfer	-	+	+
	heat storage	-	-	+
	sense of high surface heat	+	+	-
Performance related with acoustics	impact sound	-	+	+
	air born sound	-	+	+
performance related with water and moisture resistance	water resistance	-	+	+
	moisture resistance	-	+	+
Performance related with fire	inflammable	-	-	+
	continuity of structural safety -	+	+	+
	harmful gases emission	-	-	+
	keep the fire in its place	-	-	+
	keep the smoke in its place	-	-	+
Performance related with durability	resistance to chemicals	+	+	+
	resistance to biologicals	-	-	+

		+	+	+
performance related with sustainability	res. to mechanical movements	+	+	+
	CO2 footprint from the production	+	+	-
	CO2 footprint from the usage	-	-	+
	recycling	+	+	-
		9/22	15/22	19/22

Windows

The original details of the windows of historical Bosporus buildings are wooden with single glazing. The lower sash is vertical sliding and the upper sash is fixed. In some of the examples a special counter balance mechanism which is buried inside the frame exists to operate the vertical sliding lower sash. The U-value of this detail has been calculated and is approximately 5.1 W/m2K.

In table 3 performances related with original timber, and reconstruction alternative timber windows are evaluated. Most of the reconstructions use the same detail for windows on Bosporus district which is vertical sliding, wooden with double glazing. In some of the constructions both the upper and lower sashes are operable but usage of the original counter balance operating system is rather rare. Instead modern spring type operating systems are adopted in some of the details. The usage of a solar control system is also used in most of the reconstruction details. These vary largely; wooden horizontal pivoting solar shutters, figure 3 left, and the aluminium roller shutters, figure 3 right, are two largely used but significant examples of the windows of Bosporus district. The U values of these windows have been calculated, the first one without any shutters is approximately 3.3 W/m2K, the second one with wooden shutters closed is approximately 0.71 W/m2K, and the final one with aluminium shutters closed is approximately 0.31 W/m2K.

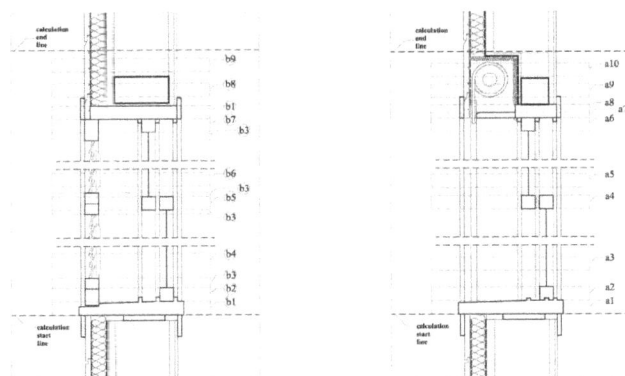

Figure 3. (left) section of the window of the reconstructions with wooden shutters; (right) section of the window of the reconstructions with aluminium roller shutters (Source: Author).

Table 3. Evaluation of the original window with reconstruction alternatives.

		original window	with wooden shutter	with aluminium roller shutter
Performance related with thermal issues	low heat transfer	-	+	+
	sense of high surface heat	-	-	-
Performance related with acoustics	impact sound	-	-	-
	air born sound	-	+	+
Perf. rel. w. water &	water resistance	+	+	+

moisture resistance	moisture resistance	-	+	+
Performance related with durability	resistance to chemicals	+	+	+
	resistance to biologicals	-	-	-
	res. to mechanical movements	-	+	+
Performance related with sustainability	CO_2 footprint of production	+	-	-
	CO_2 footprint of usage	-	+	+
	recycling	-	-	-
		3/12	7/12	7/12

When reconstruction alternatives, without any shutters, are compared with the original detail the benefit of the double-glazing is distinctive. On the other hand the effect of shutters is significant, when closed they are performing even better than the external walls. But the air space between the window and the shutter is accepted to be fully sealed as written in the related ISO 10077-2 standard (Standards International, 2012). Thus a critical test should be made to determine the exact levels.

Finally, as the sashes and frames of the original and alternative windows are kept the same any other significant performance differentiation was not determined.

Roof

The roofs of the original buildings were wooden structured. The residential usage of the attics were limited because of the crowd of studs supporting the roof. Mainly these roofs were typically consisted of tiles that stood on wooden board of 2 cm, the wooden board was being supported by rafters of 5x10cm which were standing on top of purlins of 10x10cm, purlins were supported by studs of 10x10cm, and finally the rafters, purlins, and studs are connected to each other by collar ties of 5x10cm. This detail didn't have any thermal insulation material. The air inside the attic was acting as a buffer zone for thermal issues and improved the thermal performance of the normal storeys but the attic itself was unsuitable for residential purposes. In table 4 performances related with wooden, and steel structured roofs are evaluated.

There are mainly 2 different types of details for roofs. The first detail is used if the building is wooden structured. This detail is very similar with the original detail but plenty of thermal insulation is used between the wooden rafters and the rafters are covered from below with a board, in figure 4 left, a section of these kinds of roofs may be seen. The resulting U-value of the roof has been calculated and is approximately 0.40 W/m2K. The disadvantage of this detail is about the free, usable space of the attic. There should be too many studs to support the purlins of the roof structure which minimise the total usable area of the attic.

The second roof detail is used in buildings which are R.C. or steel structured. Instead of wooden rafters, steel rafters are used which are supported by steel purlins, in figure 4 right a section of these kinds of roofs may be seen. As the steel purlins' structural capability is greater than the wooden purlins they are supported by less studs, which maximise the total usable are of the attics. Again there is plenty of thermal insulation over the rafters. The resulting U-value has been calculated and is also approximately 0.40 W/m2K.

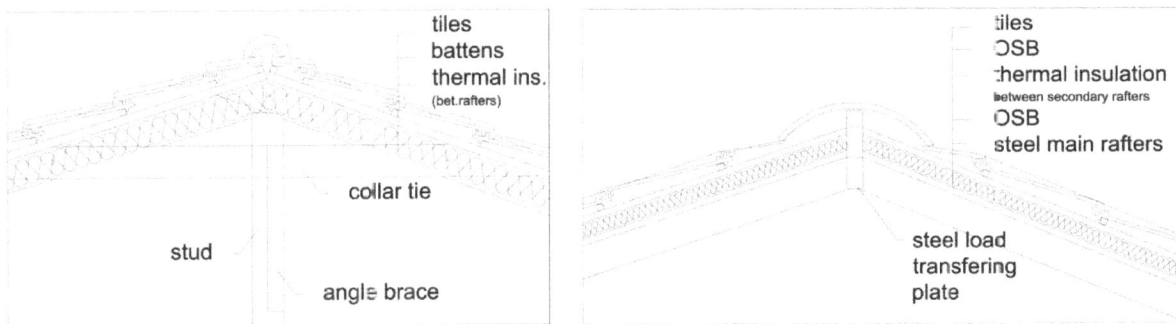

Figure 4. (left) section of the wooden roof; (right) section of the steel roof (Source: Author).

Table 4. Evaluation of the original roof with reconstruction alternatives.

		original roof	wooden roof	steel roof
Performance related with structural strength and stability	own weight	+	+	+
	earthquake	+	+	+
	impacts	+	+	+
	wind	-	+	+
Perf. related with thermal issues	low heat transfer	-	+	+
	heat storage	-	-	-
Perf.rel.w. acoustics	impact sound	-	+	+
	air born sound	-	+	+
Perf. rel. w. water & moisture resistance	water resistance	-	+	+
	moisture resistance	-	+	+
	inflammable	-	-	+
Performance related with fire	continuity of structural safety	+	+	-
	harmful gases emission	-	-	+
	keep the fire in its place	-	-	-
	keep the smoke in its place	-	-	+
Performance related with durability	resistance to chemicals	+	+	+
	resistance to biologicals	-	-	+
	res. to mechanical movements	+	+	+
performance related with sustainability	carbon footprint of production	+	+	-
	carbon footprint of usage	-	+	+
	recycling	+	+	+
		8/21	15/21	17/21

DISCUSSION

The growing demand for housing in Istanbul and the attractiveness of the Bosporus increase the demand for the restorations and reconstructions of the historical buildings. This demand brings out a contradiction with the related legislation about the Bosporus which's main aim is to stop any kind of construction in the region unless it's a historical building. The main challenge related with this legislation is the need to transform the historical buildings original details into new details for contemporary needs. These needs are both related with; the conceptual designs of the internal spaces of the buildings, and the performance requirements of the systems. For example, larger spaces are needed in the residential contemporary houses which means that there is a need for larger spans inside the buildings. Or, another example may be related with the thermal performance of the external walls, today external walls cannot be thought without insulation. It is usually too hard and inappropriate to make these kinds of changes in the existing historical buildings, but for historical buildings which do not exist anymore, these kinds of changes may be appropriate while reconstructing them. The challenges about the reconstructions may be discussed under four main topics; the structural system, the external walls, the windows, and the roofs.

Structural System

The structural system of the reconstructions of the historic buildings are demanded by the legal authorities to be the same as the original building. The idea behind this demand is to use the authentic construction techniques and to reconstruct the building as original as possible. But the original structural material may sometimes be insufficient for contemporary needs of the users. On the other hand the same legal authorities demand the building to be appropriately designed according to the legislations like structural, fire, thermal, etc. This demand contradicts with the first demand of the legal authorities and makes the contractors construct some parts of the buildings with "illegal" methods. In figure 5 left, a reconstruction example from Bosporus, dated to 2013, can be seen. The building was originally a wooden structured building. It is reconstructed with wooden structure but in order to pass larger spans the wooden structure is supported by steel beams. In figure 5 middle, a wooden truss beam of a reconstruction dated to 2012 may be seen. In figure 5 right, a fire wall of a reconstruction of another historic building may be seen, in order to realize the reconstruction the existing masonry fire wall has been demolished and a new R.C. wall has been constructed. Steel angle profiles has also been attached to support the wooden beams of the structure. In all of the examples although the original structural system material is used, the resulting buildings' structural systems do not have any authenticity which is believed to be the intention of the legal authorities.

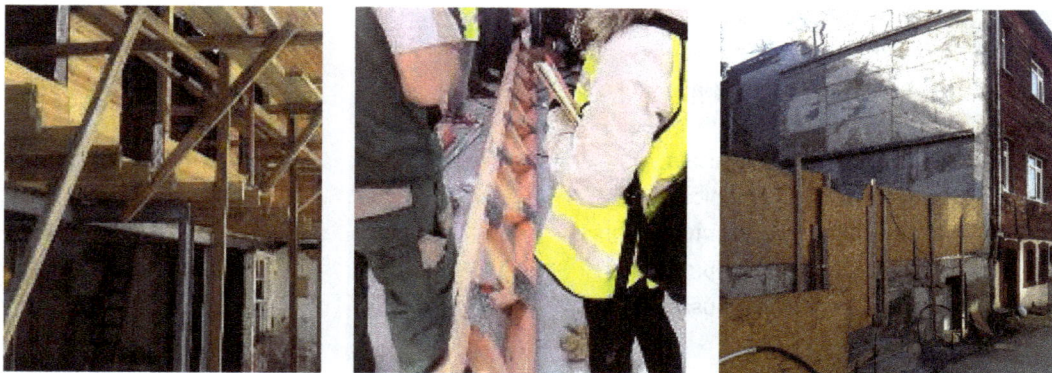

Figure 5. (left) steel beam and column supporting the wooden structure; (middle) wooden truss used as the main beam of a wooden structure; (right) R.C. firewall and anchored steel angles for supporting the wooden beams (Source: Authors).

External Walls

The external walls of the reconstructions of the historic buildings are demanded by the legal authorities to be the same as the original building. The idea behind this demand is to use the authentic construction techniques and to reconstruct the building as original as possible. But no matter what type of original external wall materials used, the performances expected from the external walls are the same and same performance improvement materials are used in reconstructions. In figure 6 left, a reconstruction of a historic building with blockwork external walls and in figure 6 middle left, a historic building with wooden stud external walls may be seen. Although the systems used are entirely different than each other the water proofing material used is the same and even in the construction process the difference cannot be distinguished. And once the reconstructions finish t is almost impossible to understand the type of the external walls.

Windows

The legal authorities let some minor changes about some parts of the windows. The main idea lies beneath this is to make the windows look like the original windows. The changes which may be realised in the windows are; the usage of double glazing, the usage of contemporary spring type opening mechanisms, operable design of the upper sashes, new locking systems, etc. The changes which are not accepted by the legal authorities are; the changes about the overall appearance of the windows, the changes about the opening type of the sashes, the usage of a solar control systems (if there wasn't one in the original building), etc. Figure 6 middle right and figure 6 right, are photographs of a window of a reconstructed building taken at different times. Figure 6 middle right is taken at the end of the construction process and figure 6 right is taken about a year later when the user added a new solar control system.

Figure 6. (left) high-density polyethylene fibre water proofing material used in a R.C. reconstruction; (middle-left) high-density polyethylene fibre water proofing material used in a wooden reconstruction (middle-right) wooden window immediately after the reconstruction, without any solar control system; (right) the same wooden window sometime after the reconstruction, with a solar control system (Source: Author).

Roofs

The roof systems of the reconstructions of the historic buildings are demanded by the legal authorities to be the same as the original building. Also the usage of the attic is accepted by the legal authorities only if it is used in the original building. On the other hand the usage of insulations in the structure of the roof system is accepted by the legal authorities which disrupts the authenticity of the roof. In figure 7 two exemplary buildings' roofs may be seen, the roof windows demonstrates that the attic is transformed to a usable space. The tiles used in figure 7 left has a special form which is manufactured by a recent manufacturer, and the tiles used in figure 7 right are concrete tiles. Both type of tiles do not represent any authenticity because originally the tiles used in these kinds of buildings were "Turkish pantile" type tiles.

Figure 7. (left) roof with roof windows and contemporary type of roof tiles; (right) air vents for ventilating roof layers and roof windows changing the overall appearance (Source: Author).

CONCLUSION AND A TENTATIVE PROPOSAL FOR RECONSTRUCTIONS OF HISTORIC BUILDINGS

The Bosporus is a unique place which has a traditional architecture. In order to protect traditional style, a special law dated to 1984 has been acted by Turkish government (Grand National Assembly, 1984). The law bans the construction of any new building along the Bosporus coast line. A construction is possible only if it is a reconstruction of an original historic building which was demolished for any kind of reasons. Since 1984 legal authorities act different legislations about these reconstructions. For example until 2005 the legislation gives permission to some of these reconstructions to be built in any structural material unless the overall appearance is the same as the original building. But after 2005 the legislation changed and the original structural material is demanded to be used in the reconstructions. The underlying reason beneath the new legislation is to protect the originality and authenticity of the Bosporus buildings both with their appearances and their construction techniques. But the practical output of this approach of the legislation makers promotes the constructers to make the reconstruction without following the approved reconstruction projects and also change the reconstructions afterwards. Addition of steel beams into wooden structured reconstructions, addition of roof windows, addition of solar control systems to the windows are some examples for these inappropriateness. This fact is a problem which is inappropriate, and which decreases the overall quality of the buildings. In the paper different common features of the reconstructions are compared and contrasted with a performance evaluation of different materials, components, elements, and/or construction techniques. This performance approach is tentatively proposed for the legal authorities to be realised for each reconstruction and decide on all kinds of materials, components, elements, and/or construction techniques considering the performance evaluation.

REFERENCES
Berge, B. (2009). *The ecology of building materials*. Amsterdam: Architectural Press.
Bougdah, H. (2009). *Environment, technology, and sustainability*. New York: Taylor & Francis.
Brock, L. (2005). D*esigning the exterior wall*. Haboken: John Wiley & Sons.
Carmody, J. et al. (2000). Residential windows: a guide to new technologies and energy performance. New York: Norton.
Cleveland, C.J. & Morris, C.G. (2009). *Dictionary of energy: expanded edition*. Amsterdam: Elsevier Science.
Cole, R., and Kernan, P. (1996). Life-cycle energy use in office buildings. *Building and Environment*, 31(4), 307-317.
Cook, M.G., (2011). *The zero-carbon house*. Ramsbury: Crowood.
Cultural and Natural Heritage Preservation Board. (2005). Decisions about the Principles for Grouping, Maintaining and Repairing the Cultural Real Estates (in Turkish). Retrieved from: http://teftis.kulturturizm.gov.tr/TR,13911/ilke-kararlari.html , access date: 13 October 2013.

Cultural and Natural Heritage Preservation Board. (2005). The Criteria for the Protection and Usage of the Urban Archeological Protected Areas (in Turkish). Retrieved from: http://teftis.kulturturizm.gov.tr/TR,13911/ilke-kararlari.html , access date: 13 October 2013.

Engel, H. (1981). *Structure systems*. Stuttgart: Deutsche Verlags.

Grand National Assembly. (1983). Law of Bosporus. *Turkish Republic Governmental Journal (in Turkish)*, 22, 970-984.

Harrison, H.W. (2000). Roofs and roofing: performance, diagnosis, maintenance, repair and the avoidance of defects. Watford: Brepress.

Hisar, A.S. (2012). Mansions of the Old Times (in Turkish), Istanbul: Yapı Kredi Publications.

Kanadoğlu, S. (2009). *Cultural and natural heritage protection law*. Ankara: Seckin Yayincilik.

Lovell, J. (2010). *Building envelopes: an integrated approach.* New York: Princeton Architectural Press.

Ministry of Environment and Urbanization. (2008). Legislation of Energy Efficiency in Buildings, *Turkish Republic Governmental Journal (in Turkish)*, 2008, 27075.

Istanbul Metropolitan Municipality. (2007). *İstanbul Building Construction Code*. İstanbul: Istanbul Metropolitan Municipality.

Istanbul Metropolitan Municipality. (2014). Duties of the Directorate of Bosporus Housing Retrieved from http://www.ibb.gov.tr/tr-tr/kurumsal/birimler/bogaziciimarmd , access date: 13 October 2013.

Rich, P. & Dean, Y. (1999). *Principles of element design.* London: Routledge.

Sakaoglu, N. (2012). What does Yalı means, does it only given to houses on the Bosporus? (in Turkish). *NTV History.* Istanbul: Dogus Yayin Grubu.

Standards British. (2000). Thermal performance of buildings – Calculation of energy use for heating Residential buildings, BS EN 832:2000, Available from: BSOL. [29 May 2015].

Standards International, (2012). ISO 10077-2, Thermal performance of windows, doors and shutters — Calculation of thermal transmittance, Part 2: Numerical method for frames

Turkcu, C. (2003). *Contemporary Structural Systems (in Turkish).* Istanbul: Birsen Yayinevi.

Yazicioglu, F. (2012). A Comparative Analysis of Embodied and Operational CO_2 Emissions from the External Wall of a Reconstructed Bosporus Mansion in Istanbul, in Hakansson, A., Höjer, M., Howlett, R.J., Jain, L.C. (Eds.) Proceedings of the 4th International Conference in Sustainability in Energy and Buildings (SEB´12), 1093-1105.

AUTHOR

Fatih Yazicioglu
Assistant Professor, PhD
Istanbul Technical University, Faculty of Architecture
fyazicioglu@gmail.com, yaziciogluf@itu.edu.tr

12

ANALYZING THE CRITICAL ROLE OF SKETCHES IN THE VISUAL TRANSFORMATION OF ARCHITECTURAL DESIGN

Farshad Helmi* and Khairul Anwar Bin Mohamed Khaidzir
Department of Architecture, Faculty Built Environment
Universiti Teknologi Malaysia

*Corresponding Author's email address: farshad_hlm@yahoo.com

Abstract
Through sketches, designers can seek and create more desirable and sustainable forms by transforming previous images through various techniques like visual additions, deletion, and modifications. Transformative skills in the form of freehand sketches appear to induce creative, explorative, open-ended environments that are conductive in dealing with the ill-structured nature of design activities. This study compares sketching and design transformative skills (DTS) between 3rd and 5th year architectural students as measured throughout the discernible levels of diagrammatic, preliminary, refinement, and detail designing. Fourteen architecture students from the University Technology Malaysia (UTM) were observed, with seven respondents each from the third and fifth year student cohorts. The objective of the observation was to capture and analyze the students' sketches as they design a gallery within the stipulated two-hour period. The research instrument included a set including an HD video camera, drawing instruments, and a brief outline of the design tasks. The Mann-Whitney test was used to determine if there were differences in design transformation activities between third and fifth year students throughout the period of observation. The results reveal significant differences in vertical move transformation between third and fifth year students within the preliminary, refinement, and detail phases of designing.

Keywords: *Sketching; design transformation skills; design expertise*

INTRODUCTION

Freehand sketching is helpful for designers because of imagining and recognising many drawing alternatives. Sketch has the crucial role of supporting the mind in converting descriptive information into depiction. Fish (1996) mentions that sketch aids artist to think and support short-term memory. Sketch aids designers in considering problems and is a beneficial technique for communicating ideas to others (Myers et al, 2008). Some of the research in design recognizes that sketching is an effective instrument for conceptual designing. In addition, it can serve as a storage solution as external memory and appear to be important in understanding conflicts and possibilities (Akin, 1978). Therefore, sketching has an important role in design training and it can be beneficial for design development and design creativity.

Transformation is the mechanism that shows the way new designs are generated from unambiguous representations and the prevailing products. Moreover, this is the design that seeks to create desirable and sustainable changes in form (Tovey, Porter, and Newman, 2003). In order to transform descriptions into depictions, the designer employs a set of quick sketches. In this manner, images are generated in mind by sketches, by which the embodied themes in the design are developed. Sequentially, this directs the designer to transform the former image through additions, deletions, and modifications (Tovey et al, 2003). Indeed, transformation moves from unstructured drawing to further detailed and precise illustrated representations.

This study uses retrospective protocol analysis to compare design transformation in four levels of detailing: diagrammatic, preliminary, refinement, and detail designing between third and fifth year students. The contribution of this study is to aid in understanding the difference of design transformation between two levels of design expertise. It is expected that this study will be beneficial for design education and design learning.

LITERATURE REVIEW

Transformation is the capability of pattern modification in images and is significant in design generation. This transformation occurs in different context, which will be explained later as a decomposing process. Sketching manipulation takes place for transforming image, situation, and drawing types that result in evaluating design solution (Do et al, 2000).

Transformation

Design transformation is the progression from unstructured form to structured form which occurs for creating, modifying, and developing design elements and the design idea. Goel (1995) argued that design is the process from ill-defined problems to the well-defined design problems. It consists of some moves that start from the preliminary phase (unstructured sketch) and the refinement phase of design to detail design (explicit and precise design). Goel (1994) stated that lateral and vertical is transformation, whereas duplication is repeating. A lateral transformation is identified as "movement... from one idea to a slightly different idea". They are essential for broadening the problem space and the assessment and improvement of kernel ideas. A vertical transformation is identified as "movement... from one idea to a more detailed version of the same idea". It causes the problem space to deepen. Lateral transformations mostly take place in the initial design stages and are related to unstructured drawing while vertical transformations take place throughout the refinement and detailed design stages and are related to more precise and detailed design.

Van der Lugt (2000) investigated features of design transformation that occur in idea links. He defined ideas as three subclasses in a link: supplementary, modification, and tangential links. The supplementary link shows auxiliary and small change on the same version of the idea; the modification link relates to changes in the structure of ideas, however keeping the current line of thought; the tangential link indicates a radical and fundamental change from the earlier idea. He describes that a higher tangential link showed that design ideas have rich novelty whereas a higher supplementary link and modification link indicated development in the idea. Van der Lugt (2000) also mentioned that a creative process consists of a balance among link types. Similarly, Rodgers, Green, and McGown (2000) mention that the balance between vertical and lateral transformation results in good design.

Abdelmohsen and Do (2007) investigate the concept development of two Ph.D. students that had two and six years of professional experience. They evaluate the development of the concept based on vertical and lateral transformation in three seasons: creating design options in schematic design, developing and refining design options, and detailing the revised design. A protocol analysis was used for studying concept development in the schematic and refined stages of design. Moreover, they decompose the design process to three meaningful strokes, namely: transferred, blocked, or added. Abdelmohsen and Do (2007) extended the notion of design transformation by defining processes described as vertical promotion and lateral promotion and cross propagation. Indeed, they evalute design transformation by meaningful strokes in the macro level.

Differences between two levels of design expertise

Different studies have emerged that offer contradictory finding about differences in design expertise. Suwa and Tversky (1997) compare the design thoughts of two groups, advanced students and practicing architects, during the sketching of a museum. A retrospective protocol analysis was considered for this issue. The authors observed that the whole process of design consisted of two groups of segments, the alone segment which they named the isolated segment and the contiguous segment which is set in one block that the authors named the dependency chunk. The isolated segment and the initial segment of the dependency block shows that designers focus on previous thought and shift to an alternative topic, item, or space. Suwa and Tversky (1997) name these segments by means of focus-shift that corresponds to lateral transformation; in addition, with the exception of the initial segment, they name further segments located in the dependency chunk as continuing segments that relate to vertical transformation. The authors concluded that practicing architects used dependency chunk longer and more than advanced students.

Kavakli and Gero (2001) investigated the cognitive processes of novice and expert designers in which the expert designers produced 7 alternatives whereas the novice students had 2, thus they conclude that alternative interpretations' perception and spatial relations' organization may consume more time for the novice than expert designers (Kavakli and Gero, 2001). They described that the main difference in their sketches is that there was more intensity in the representation of design ideas as seen in the expert's design alternatives (Kavakli and Gero, 2001). Moeover, Atman et al (2005) gathered verbal protocols from first year engineering students (freshman) with fourth year engineering students (senior) while they worked on two design problems. They define that, in contrast to the senior students, freshmen considered less alternative solutions, gathered less information, and transitioned less frequently between types of design activities. By comparing literature, it is obvious that the "expert is more productive than the novice" based on the quantities of alternatives and pages created; practicing architects utilized vertical transformation more and longer than advanced students.

Sketch
In the primary phase of the design process, sketch has a crucial role among the traditional mediums and is the elementary depictive action that is performed by designers during the design process. Garner (1990) mentions that sketching fundamentally affects the development, creation, evaluation, and distribution of ideas. Moreover, Goel (1995) suggests that being "syntactically" and "semantically" unclear and ambiguous, the sketches influence the heuristic, creative, open-ended stage of problem-solving. Some researchers like Fish and Scrivener (1990), Goel (1995), and Goldschmidt (1991) came to the conclusion that rough and untidy sketching allows the designer to work quickly, suspending judgment on polished features. Moreover, it could help in generating new ideas. Purcell and Gero (1998) state that in design perception research, a considerable number of studies have focused on the roles of sketches in the conceptual design process and their relationship to designer's cognition. Indeed, for simplifying the existing ideas and developing new ones, sketching can be helpful.

Decomposing Process
Several attempts have been made to decompose the whole design process into three different components for analysing and measuring it: context, chunk, and move (Figure 1).

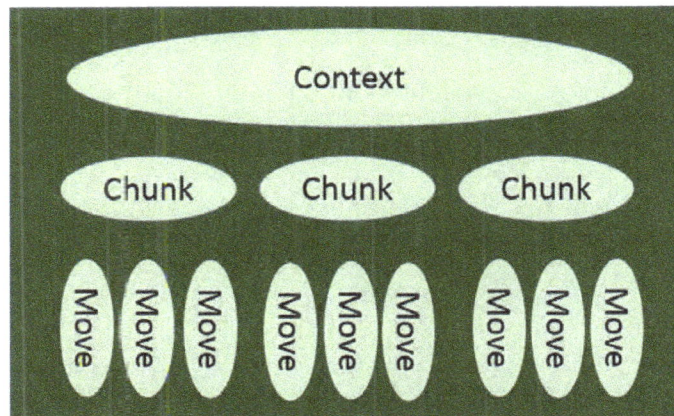

Figure 1. Decomposing design process to context, chunk, and move

Context

The Ddsigners change their focus in design and refine the domain knowledge by the context of sketch (Cai, Do, and Zimring, 2010). Gross and Do (1996) argue a similar drawing symbol might have a different sense in a different context. Do et al (2000) stated that transformation can occur in design and context trough manipulating shape and changing drawing types and viewpoints. Previous studies have classified context in the design process based on design development, the level of abstraction, and presentation types.

The first type of context is classified based on design development, Goel (1994, 1995, and 2014) categorized the development of design into the four subcategories: (1) problem-structuring, which arranges the problem, (2) preliminary-design, which creates some solution options and idea cores, (3) refinement-design, which improves the current sketch by transformation, and (4) detail-design, which presents the design product. Similarly, Abdelmohsen and Do (2007) classified seasons of design into three phases. In the first phase, several solution options are created in schematic plan drawings. In the second phase, designers refine and improve options. In the third phase, they improve refined drawing to the product of design and organization elements (Table 1). Although ciagram has an important role in design thinking and sketching types, design development of Goel do not include diagrams. For Goel, diagrams are more related to function, whereas preliminary design is related to form conceptual design. It seems that it may however be more useful to add diagram to design development.

The second type of context is the abstraction level. Designers use abstract diagrams and unstructured forms in early phases of the design process, while they utilize detailed and structured representations in later phases of the design process (Purcell and Gero, 1998). Fish and Scrivener (1990) categorize the element of pictorial representation from description to spatial depiction and argue that sketch has an essential role in supporting the mind by interpreting the "descriptive propositional information" to depiction. Goel (1995) in "Sketch of Thought" mentioned that the design process contains some movement from ambiguity and vague shape, which is important in the early phase of design to more structure form in detailed design. Consequently, this is a process of developing from unclear sketch to detailed form; he notes that the design transformation process moves from abstraction level to convention document. He (1995) describes drawings as "external symbol systems for representing the real world artifacts". He (1995) familiarizes some drawing from graphic thinking of Laseau (2001) and classifies them as the symbolic system. Goel (1995) lists them as bubble diagram, layout diagram, conceptual sketch, first sketch of floor plan, schematic of floor plan, quick freehand perspective and some detail. Cai et al (2010) classified stimuli for inspiration sources regarding levels of abstraction demonstrated different contexts These stimuli are adapted from "Frank Lloyd Wright's Robie

House". They categorize these stimuli as "keyword, diagram, plan, sketch rendering, and precedent photo" (Table 1).

Last, the third type of context is presentation. Do et al (2000) and Bar-Eli (2013) pointed out several types of projection (presentation types) like elevation, plan, elevation, section, and perspective (Table 1).

Table 1. Three different context in design process

Context	Author	levels
Design development	Goel (1994, 1995, 2014)	(1): Problem-structuring, (2): Preliminary-design, (3): Refinement-design and (4): Detail-design
	Abdelmohsen[1] and Do[1] (2007)	(1): schematic design phase, (2): refine design phase and (3): detail design phase
Abstraction level	(Purcell & Gero, 1998)	Unstructured form to structure representation
	J. Fish and Scrivener (1990)	Description to spatial depiction
	Goel (1995)	Bubble diagram, layout diagram, conceptual sketch, first sketch of floor plan, schematic of floor plan, quick freehand perspective and some detail.
Presentation types	Do et al. (2000) and Bar-Eli (2013)	Plan, section, elevation, and perspective

Chunk

Goldschmidt (1992) recognized that chunk consists of some moves and the relationships between them. She defined chunk as "the block of links among successive moves that link exclusively among themselves and [are] barely interconnected with other moves". Goldschmidt (1992) restricts chunk according to design moves with the large number, whereas Suwa and Tversky (1997) structured them based on segments with the smaller number. Chunk is often used in linkography, as opposed to other methods.

Move

Goldschmidt (1990) decomposes the design process into small parts of "design moves" by using linkography. Goldschmidt defined a movement as "a step, an act, an operation, which transforms the design situation relative to the state in which it was prior to that move" (Goldschmidt, 1995b) or can be separate as "the smallest coherent operation detectable in design activity" (Goldschmidt, 1992). Goldschmidt (1995a) mentions that considering design move can result in analyzing and assessing the design process. Moreover, Goel (1995) defined three design movements: vertical and lateral, which are transforming, and duplication, which is repeating.

It seems that design development of Goel is more relevant to design transformation and idea development. However, diagram plays an essential role in design thinking and the lack of diagram in the previous classification of levels of detailing can be seen. Therefore, the current study improves the design development model of Goel (1994) as design transformation model and categorizes it based on four levels of detailing namely, diagram, preliminary design, refinement design, and detail design. The current study aims to measure design transformation based on movements in levels of detailing (design development).

METHOD

Data Gathering
Most design researchers used protocol analysis to measure cognitive activity that take place in designing and the process of design. The protocol was divided into two subclasses: concurrent protocols and retrospective protocols (Ericsson and Simon, 1993). Concurrent protocols are obtained from the explanation of a designer's thinking during their sketching. Retrospective protocols are obtained from the explanation of a designer's thinking after they completed their sketching. Design researchers used two types of protocol based on the nature of the problem in research design. Process-oriented design studies associated with concurrent protocols, whereas content-oriented design studies related to retrospective protocols (Dorst and Dijkhuis, 1995). Design researchers did not find any research methodology that was suitable for different goals and diverse situations (Goldschmidt and Weil, 1998). The think-aloud protocol is not used in this study because previous studies suggested that talking aloud concurrently may limit the perception of the participant during their drawing activities (Ericsson and Simon, 1993). This effect may be a weakness in our study since our aim was to investigate transformation difference between two groups. We interviewed the participants after they finished their design task as a retrospective protocol. We observed 14 undergraduate architecture students of the University of Technology Malaysia in separate individual design activities. All of the students worked on designing a gallery of seashell and stone.

Segment
Previous design researchers decomposed the verbal protocol into small parts as segmentation. Segmentation is defined according to different events. The first is verbalization actions such as intonations, pauses, and syntactic signs for complete sentences and phrases (Ericsson and Simon, 1993; Gero and Mc Neill, 1998; Goldschmidt, 1991). The second is defined according to the "subject's intention" (Goldschmidt, 1991; Suwa and Tversky, 1997). For instance, Goldschmidt (1991) determined a segment as design move, which is defined as "an act of reasoning which presents a coherent proposition pertaining to an entity that is being designed". A change in the designer's thought contents, their action, and their intention in a subject for the sign for the start and the end of the new segment. Therefore, one segment sometimes contains many sentences and sometimes only one. Moreover, Chiu (2003) analyzes design activities of designers in equal time sequence. We use the latter approach, which means that segmentation in the current study employed equal segments for measuring design transformation in ordinal scale. Since two movements have different time periods, this study divided design process into 30-second equal segments. By considering equal segments, we can measure design transformation more exactly in ordinal scale. It means that one movement can consist of one segment or more than one segment.

Coding
In design studies, coding schemes are needed for defining different action categories in protocol analysis. There are a variety of developed coding schemes depending on the purpose and the scope of every study. The retrospective protocol analysis method applied in this study is based on a content-oriented approach. This section proposes a unique method for tabulating transformation activities. It follows earlier suggestions that each complete segment is encoded with relevant attributes (in bracket) under the following three categories: (1) transformation type

(lateral or vertical), (2) presentation types (ground floor, first floor, section, elevation, and perspective), and (3) levels of detailing (diagram, preliminary design, refinement design, and detail design). These attributes then populate a table referred to as the matrix of design transformation. The tables were created using Microsoft Excel. The following sections highlight the basis for the current tabulation format (Table 2).

Table 2. Sub codes of presentation types, transformation types and levels of detailing in DTM

PRESENTATION TYPES	TRANSFORMATION TYPES	PHASES OF DETAILING
GF: Ground floor	V: Vertical	DI: Diagram
FF: First floor	L: Lateral	PR: Preliminary design
SE: Section	0: No movement	RE: Refinement design
EI: Elevation	O.M: Another movement that is not transformation like duplication	DE: Detailing design
PE: Perspective		

Design Transformation

As mentioned before, Goel (1995) used the term transformation in describing the movements of ideas: a lateral transformation indicates a "shift from one idea to a different idea", while vertical transformation suggests a detailed development of the same idea.

Levels of Detailing

This study updates the design development model of (Goel, 1994) as a design transformation model and categorizes it based on four levels of detailing namely, diagram, preliminary design, refinement design, and detail design. In the diagrammatic phase of the design process, designers evaluate functions, position, and relationships between spaces as in a bubble diagram. That is consistent with unstructured drawing and abstract schematic. In the preliminary phase of the design process, designers try to generate shape, concept, and kernel ideas. They keep alternatives open and prospects broad. That is the early effort to produce shape. In refinement design, designers develop and revise kernel idea or early generated design idea. In detailing design, designers take the final shape of the form of the idea that consists of precise and structured forms of dimensioned drawing and straight line (Table 3).

Design Protocol Measurement

Khaidzir and Lawson (2013) propose a Cognitive Interaction Matrix (CIM) framework for examining the intricate nature of design studio interactions. CIM illustrates encoded cognitive segments for each design tutorial meeting recorded for the aim of the study. The CIM multi-coding framework offers an inclusive and systematic cognitive description for each protocol segment. This study uses a multi-coding matrix in a different content, as a design transformation matrix (DTM) to code and insert data. This matrix converts qualitative data to quantitative data (Table 4).

Table 3. New model of design transformation in levels of detailing based on Goel's framework (Goel, 1994).

Design oriented	Design development	Feature	Sketch	Levels of detailing

Unstructured	Core Of Idea and New Generation	In the diagrammatic phase of the design process, designers evaluate functions, position, and relationships between spaces such in bubble diagram. That is consistent with unstructured drawing and abstract schematic.		Diagram
		In the preliminary phase of the design process, designers try to generate shape, concept, and kernel idea; they keep alternatives open and prospects broad. That is the early effort to produce shape		Preliminary Design
	Idea Development	Designers develop and revise kernel ideas or early generated design idea in refinement design.		Refinement Design
Structure	Fixation and Detailing	In detailing design, designers take the final shape of the form of the idea that consists of precise and structured forms of dimensioned drawing and straight line.		Detail Design

Table 4. Design transformation matrix (DTM), organized based on the equal segment.

Segment No	Time Duration	Sheet	Presentation Types	Transformation Types	Phases Of Detailing
59	0:29:00	5	GF	L	PR
60	0:29:30	5	GF	L	PR
61	0:30:00	5	GF	L	PR
62	0:30:30	5	GF	L	PR
63	0:31:00	5	0	0	0
64	0:31:30	5	GF	L	PR
65	0:32:00	5	GF	V	PR
66	0:32:30	5	GF	V	PR
67	0:33:00	5	GF	O-M	PR
68	0:33:30	5	0	0	0
69	0:34:00	5	GF	V	PR
70	0:34:30	5	GF	V	PR

Scope

The University of Technology Malaysia has two architecture programmes for undergraduate architecture students, the old one is five years, and the new one is three years. This study compares undergraduate students in the two groups. As such, 14 undergraduate architecture students at the university technology Malaysia (UTM) are involved in this experiment, 7 third and 7 fifth year students. They were asked to complete a design task in two hours. The research instruments include a video camera, a set of drawing tools, and the design task.

Analysis

All statistics are analyzed using the Mann-Whitney U-tests (Clark-Carter, 1997) through SPSS statistical analysis software. Mann-Whitney U-tests are used to test the differences between two independent samples. (Mann-Whitney U is the non-parametric equivalent of a t-test and compares the ranked scores of the two groups).

FINDINGS AND DISCUSSION

In this part, the null hypothesis is that the period of transformation of both groups is equal in four levels of detailing, so if the result is significant, the null hypothesis is rejected. The Wilcoxon-Mann-Whitney test was applied on the ranked using the period of transformation between third and fifth year students. Table 5 shows the periods of transformation that were produced by third and fifth year students. For this experiment, the Wilcoxon-Mann-Whitney test was chosen to establish if third and fifth years students hold similar (null hypothesis) or different (alternative hypothesis) statistical distribution in design transformation. It can be seen from the table, in the diagrammatic phase of detailing, the vertical (P=0.40) and lateral (P=0.72) p-values were beyond the 0.05 level. Therefore, the Mann-Whitney test failed to establish significance beyond the 0.05 level in the difference of transformation between two groups in the diagram. In preliminary and detail design, the p-values are 0.015 and 0.034 respectively. This indicates that third year students spend more time for vertical transformation in these two levels of detailing.

On the other hand, fifth year students use vertical and lateral more than third year students in refinement design; this is indicated by p-values below 0.05, and the significance of vertical and lateral is 0.015 and 0.045 respectively. Thus, although third year students have more activity in idea development in preliminary and detail design, fifth year students spend more time during idea development and modification of their idea in refinement design.

Table 5. Significant of design transformation difference in 4 levels of detailing

Levels Of Detailing	Transformation	Significant	Ranks	
			3th Years	5th Years
Diagram	Vertical	0.40	46	59
	Lateral	0.72	50	55
Preliminary Design	Vertical	0.015	71.5	33.5
	Lateral	0.14	64	41
Refinement Design	Vertical	0.015	33.5	71.5
	Lateral	0.045	37.5	67.5
Detail design	Vertical	0.034	69	36
	Lateral	0.21	43	62

CONCLUSION

The aim of this study was to evaluate differences in design transformation between third and fifth-year undergraduate students in 4 levels of detailing. Findings indicate some significant difference of transformation between the two groups. By considering time segments, it is obvious that third-year student use vertical transformation longer than fifth years in preliminary design and detail design, while the third year students utilize vertical and lateral for less time than fifth year students in refinement design. Thus, third year students spend more time generating each alternative solution and kernel idea than do fifth year students in preliminary design. On the other hand, fifth year students utilize vertical (convergence thinking) and lateral (divergence thinking) more than third year students in refinement design. In other words, fifth year students develop in-depth and revise kernel ideas more in this phase. Finally in the detailing phase, third year students develop their idea more than fifth year students. Therefore, fifth year students fix their design in refinement design more than third year students. This finding indicates how students use different Design Transformation Skills (DTS) among levels of design detailing. The results of this study are expected to be beneficial for the distinction of design expertise levels. We plan to compare further the result of equal segments (ordinal scale) with real time (ratio scale) to understand differences between them.

ACKNOWLEDGEMENT

This study is funded by Universiti Teknologi Malaysia, Skudai through the Teaching Development Grant (Dana Pembangunan Pengajaran, DPP 2015) entitled 'Peer Instruction in Virtual Collaborative Environment of Studio Based Learning' (R.J130000.7721.4J165).

REFERENCES

Abdelmohsen, S., and Do. E.Y.-L. (2007). *Tracking design development through decomposing sketching processes*. Proceedings of the International Association of Societies of Design Research (IASDR 2007), Emerging Trends in Design Research, Hong Kong Polytechnic University School of Design, Hong Kong.

Akin, O. (1978). How do architects design. In. J.C. Latombe (Ed.), *Artificial Intelligence and Pattern Recognition in Computer Aided Design* (65-119). New York, N.Y.: North-Holland Publishing.

Atman, C.J., Cardella, M.E., Turns, J., and Adams, R. (2005). Comparing freshman and senior engineering design processes: An in-depth follow-up study. *Design Studies, 26* (4), 325-357.

Bar-Eli, S. (2013). Sketching profiles: Awareness to individual differences in sketching as a means of enhancing design solution development. *Design studies, 34* (4), 472-493.

Cai, H., Do, E.Y.L., and Zimring, C.M. (2010). Extended linkography and distance graph in design evaluation: An empirical study of the dual effects of inspiration sources in creative design. *Design studies, 31* (2), 143-168.

Chiu, M.L. (2003). Design moves in situated design with case-based reasoning. *Design studies, 24* (1), 1-25.

Clark-Carter, D. (1997). *Doing quantitative psychological research: From design to report*. London, U.K.: Psychology Press.

Do, E.Y.L., Gross, M.D., Neiman, B., and Zimring, C. (2000). Intentions in and relations among design drawings. *Design studies, 21* (5), 483-503.

Dorst, K., and Dijkhuis, J. (1995). Comparing paradigms for describing design activity. *Design studies, 16* (2), 261-274.

Ericsson, K., and Simon, H. (1993). *Protocol analysis: Verbal reports as data* (Revised Edition). Cambridge, M.A.: MIT Press.

Fish, J., and Scrivener, S. (1990). Amplifying the mind's eye: Sketching and visual cognition. *Leonardo*, 117-126.

Fish, J.C. (1996). *How sketches work: A cognitive theory for improved system design* (Doctoral dissertation). Loughborough University of Technology, Loughborough, U.K..

Garner, S.W. (1990). Drawing and designing: the case for reappraisal. *Journal of Art and Design Education, 9* (1), 39-55.

Gero, J.S., and Mc Neill, T. (1998). An approach to the analysis of design protocols. *Design studies, 19* (1), 21-61.

Goel, V. (1994). A comparison of design and nondesign problem spaces. *Artificial Intelligence in Engineering, 9* (1), 53-72.

Goel, V. (1995). *Sketches of thought*. Cambridge, M.A.:MIT Press.

Goel, V. (2014). Creative brains: Designing in the real world. *Frontiers in human neuroscience, 8*.

Goldschmidt, G. (1990, April). *Linkography: Assessing design productivity*. Paper presented at the Cyberbetics and System'90, Proceedings of the Tenth European Meeting on Cybernetics and Systems Research, Vienna, Austria. Vienna, Austria: World Scientific Pub Co Inc

Goldschmidt, G. (1991). The dialectics of sketching. *Creativity Research Journal, 4* (2), 123-143.
Goldschmidt, G. (1992). Criteria for design evaluation: A process-oriented paradigm. *Evaluating and Predicting Design Performance*, 67-79.
Goldschmidt, G. (1995a). The designer as a team of one. *Design studies, 16* (2), 189-209.
Goldschmidt, G. (1995b). Visual displays for design: Imagery, analogy and databases of visual images. *Visual Databases in Architecture*, 53-74.
Goldschmidt, G., and Weil, M. (1998). Contents and structure in design reasoning. *Design Issues*, 85-100.
Gross, M.D., and Do, E.Y.L. (1996). *Demonstrating the electronic cocktail napkin: A paper-like interface for early design.* Paper presented at the Conference Companion on Human Factors in Computing Systems.
Kavakli, M., and Gero, J.S. (2001). Sketching as mental imagery processing. *Design studies, 22* (4), 347-364.
Khaidzir, K.A.M., and Lawson, B. (2013). The cognitive construct of design conversation. *Research in Engineering Design, 24* (4), 331-347.
Laseau, P. (2001). *Graphic thinking for architects and designers.* Hoboken, N.J.: John Wiley & Sons.
Myers, B., Park, S.Y., Nakano, Y., Mueller, G., and Ko, A. (2008). *How designers design and program interactive behaviors.* Paper presented at the Visual Languages and Human-Centric Computing, 2008. VL/HCC 2008.
Purcell, A., and Gero, J.S. (1998). Drawings and the design process: A review of protocol studies in design and other disciplines and related research in cognitive psychology. *Design studies, 19* (4), 389-430.
Rodgers, P., Green, G., and McGown, A. (2000). Using concept sketches to track design progress. *Design studies, 21* (5), 451-464.
Suwa, M., and Tversky, B. (1997). What do architects and students perceive in their design sketches? A protocol analysis. *Design studies, 18* (4), 385-403.
Tovey, M., Porter, S., and Newman, R. (2003). Sketching, concept development and automotive design. *Design studies, 24* (2), 135-153.
Van der Lugt, R. (2000). Developing a graphic tool for creative problem solving in design groups. *Design studies, 21* (5), 505-522.

AUTHORS

Farshad Helmi

Ph.D. Student of Architecture
Department of Architecture, Universiti Teknologi, Malaysia.
Email address: *hfarshad2@live.utm.my*

Khairul Anwar Mohamed Bin Khaidzir

Senior Lecturer
Department of Architecture, Universiti Teknologi, Malaysia.
Email address: *b-anwar@utm.my*

Permissions

All chapters in this book were first published in IJAR, by ArchNet-IJAR; hereby published with permission under the Creative Commons Attribution License or equivalent. Every chapter published in this book has been scrutinized by our experts. Their significance has been extensively debated. The topics covered herein carry significant findings which will fuel the growth of the discipline. They may even be implemented as practical applications or may be referred to as a beginning point for another development.

The contributors of this book come from diverse backgrounds, making this book a truly international effort. This book will bring forth new frontiers with its revolutionizing research information and detailed analysis of the nascent developments around the world.

We would like to thank all the contributing authors for lending their expertise to make the book truly unique. They have played a crucial role in the development of this book. Without their invaluable contributions this book wouldn't have been possible. They have made vital efforts to compile up to date information on the varied aspects of this subject to make this book a valuable addition to the collection of many professionals and students.

This book was conceptualized with the vision of imparting up-to-date information and advanced data in this field. To ensure the same, a matchless editorial board was set up. Every individual on the board went through rigorous rounds of assessment to prove their worth. After which they invested a large part of their time researching and compiling the most relevant data for our readers.

The editorial board has been involved in producing this book since its inception. They have spent rigorous hours researching and exploring the diverse topics which have resulted in the successful publishing of this book. They have passed on their knowledge of decades through this book. To expedite this challenging task, the publisher supported the team at every step. A small team of assistant editors was also appointed to further simplify the editing procedure and attain best results for the readers.

Apart from the editorial board, the designing team has also invested a significant amount of their time in understanding the subject and creating the most relevant covers. They scrutinized every image to scout for the most suitable representation of the subject and create an appropriate cover for the book.

The publishing team has been an ardent support to the editorial, designing and production team. Their endless efforts to recruit the best for this project, has resulted in the accomplishment of this book. They are a veteran in the field of academics and their pool of knowledge is as vast as their experience in printing. Their expertise and guidance has proved useful at every step. Their uncompromising quality standards have made this book an exceptional effort. Their encouragement from time to time has been an inspiration for everyone.

The publisher and the editorial board hope that this book will prove to be a valuable piece of knowledge for researchers, students, practitioners and scholars across the globe.

List of Contributors

Seyed Reza Hosseini Raviz, Ali Nik Eteghad, Ezequiel Uson Guardiola and Antonio Armesto Aira
Department of Architectural Projects (DPA), Polytechnic University of Catalonia (UPC), Barcelona, Spain

Emine Özen Eyüce
Bahcesehir University, İstanbul

Burak Pak
University of Leuven, Faculty of Architecture Brussels, Belgium

Renato Capozzi, Adelina Picone and Federica Visconti
DiARC_Department of Architecture, University of Naples "Federico II", Italy

Sabir Nu'Man
Ashghal Public Works Authority, Doha, Qatar

Evawani Ellisa
Universitas Indonesia, Kampus UI Depok, 16424 Indonesia

Amy Huber
Florida State University, Tallahassee, Florida, USA

Karen Kim and Edward Steinfeld
Center for Inclusive Design and Environmental Access, University at Buffalo, The State University of New York, Buffalo, NY, USA

Naif A. Haddad
Department of Conservation Science, Queen Rania's Faculty of Tourism and Heritage, Hashemite University, Jordan

Fatima Y. Jalboosh
Private Practice, Amman, Jordan

Leen A. Fakhoury
School of Architecture and Built Environment, German Jordanian University, Jordan

Romel Ghrayib
Department of Antiquities, Zarqa, Jordan

Naiana M. John, Antônio T. da Luz Reis, Márcia A. de Lima, Maria Cristina Dias Lay
Federal University of Rio Grande do Sul, Porto Alegre, Brazil

Faezeh Nabavi and Yahaya Ahmad
Department of Architecture, Faculty of Built Environment, University of Malaya, 50603 Kuala Lumpur, Malaysia

Farshad Helmi and Khairul Anwar Bin Mohamed Khaidzir
Department of Architecture, Faculty Built Environment Universiti Teknologi Malaysia

Index